Thermal Inertia in
ENERGY EFFICIENT
BUILDING ENVELOPES

Thermal Inertia in
ENERGY EFFICIENT
BUILDING ENVELOPES

FRANCESCA STAZI
Polytechnic University of Marche, Ancona, Italy

Butterworth-Heinemann
An imprint of Elsevier

Butterworth-Heinemann is an imprint of Elsevier
The Boulevard, Langford Lane, Kidlington, Oxford OX5 1GB, United Kingdom
50 Hampshire Street, 5th Floor, Cambridge, MA 02139, United States

Notices
Knowledge and best practice in this field are constantly changing. As new research and
experience broaden our understanding, changes in research methods, professional practices, or
medical treatment may become necessary.

Practitioners and researchers must always rely on their own experience and knowledge in
evaluating and using any information, methods, compounds, or experiments described herein.
In using such information or methods they should be mindful of their own safety and the safety
of others, including parties for whom they have a professional responsibility.

To the fullest extent of the law, neither the Publisher nor the authors, contributors, or editors,
assume any liability for any injury and/or damage to persons or property as a matter of products
liability, negligence or otherwise, or from any use or operation of any methods, products,
instructions, or ideas contained in the material herein.

British Library Cataloguing-in-Publication Data
A catalogue record for this book is available from the British Library

Library of Congress Cataloging-in-Publication Data
A catalog record for this book is available from the Library of Congress

ISBN: 978-0-12-813970-7

For Information on all Butterworth-Heinemann publications
visit our website at https://www.elsevier.com/books-and-journals

  Working together
to grow libraries in
developing countries

www.elsevier.com • www.bookaid.org

Publisher: Mathew Deans
Acquisitions Editor: Ken McCombs
Editorial project manager: Charlotte Kent
Production Project Manager: Paul Prasad Chandramohan
Cover Designer: Mark Rogers

Typeset by MPS Limited, Chennai, India

I wish to dedicate this book to my father
who taught me that the respect of others must be earned every day

CONTENTS

BIOGRAPHY

Francesca Stazi, PhD, is an associate professor at the Polytechnic University of Marche, Italy.

She carries out experimental and numerical research activities in the field of Building Science and Technology. The aim is to optimize the building envelope in terms of energy saving, thermal comfort, environmental sustainability, and durability of the components. The researches cover new and existing envelopes, ventilated facades, and passive solar systems. The acquired knowledge was applied in the patenting of two industrial inventions, an innovative ventilated thermal insulation and a GFRP frame for windows.

The results of the studies are reported in 65 publications, including 25 papers on international ISI Journals.

She is a reviewer for various international ISI Journals.

PREFACE

Perhaps the most uncertain and least well-understood aspect of the design of building envelope is the prediction of its dynamic behavior in relation to the indoor and outdoor variations. The difficulty to quantify the non-linear processes involved in the interaction between envelope and interior/exterior climate is a common cause of discomfort problems for the scarce exploitation of the thermal inertia benefices. Yet research into the dynamic properties of the mass and its interaction with insulation has been actively pursued for the past 40 years. However, the debate among researchers is still open since the optimal envelope solution varies based on the considered aspect between energy saving, comfort, environmental impact, global costs, and climates, and could depend on specific building features and on the highly variable user behavior. In the last decades the increasing of the insulation standard, the sophistication of the equipment available, and the developments of new constructive techniques have deeply changed the building envelope, simultaneously increasing the comfort levels expected by the occupants. These developments were not followed by a corresponding creation of a technical culture for a building envelope suited to the specific climate, giving rise to new discomfort problems. Moreover much of the information and many of the analytic techniques, which are well known to researchers, are not used by the professionals. Often the practical significance of the researches based on parameter optimization, e.g., decrement factor and time lag, has not been fully explored or explained. Designers rely on the simplified procedures contained in national regulations that simplify the complex problems since based on stationary simulations.

This book tries to fill the gap between the most advanced research on envelope optimization and the practical choice by designers of building envelopes. It is an attempt to provide researchers and practising engineers with simple and quantitative information on the performance of different kinds of building envelope in various climates and building features. This attempt is made by:

- giving experimental and numerical data on consumptions, comfort levels, and environmental/economic costs for a great number of envelopes in different climates and usage conditions. The presented data are very reliable since all the simulations are based on models calibrated through experimental measures.

- giving simple methods to provide the comfort levels for every envelope freely chosen by the readers (according to EN ISO 13786:2008). An entire chapter of the book is devoted to the explanation of the consequence of choosing one dynamic parameter than another. An appendix at the end of the book reports the calculation details for each presented parameter.
- giving indications on existing envelope optimization. Simple solutions to optimize existing envelopes are drawn and suggestions on how to behave regarding existing insulations are also included.

The author believes that the presented large collection of experimental data on all the envelope typologies is the most interesting aspect of the book, given that it is very rare in the international literature, it comes from 13 years of research and scientific publications by the author, and from extensive monitoring campaigns on more than 30 case studies. This work reports novel results and proposals also based on the reprocessing and reinterpretation (in practical sense) of data previously published in international academic field with other authors, which are hereafter acknowledged. The transversal evaluation of the great data collection on which the book is based is the only possible approach to give an answer to following open questions: is the dynamic behavior of the mass still possible in the new energy efficient superinsulated envelopes? Which are the design features that configure a climatic-"adaptive" envelope instead of a stationary one?

It is hoped that this book will provide envelope designers not only with useful information to select their optimal configuration but also with a better understanding of the dynamic behavior of the building envelope.

Francesca Stazi

Ancona, 2017

ACKNOWLEDGMENTS

I wish to thank Prof. Costanzo Di Perna for his contribution on understanding various aspects of the complex envelope dynamic behavior. He was my co-author in many papers on thermal inertia. Thanks to Prof. Placido Munafò for his valuable advice and for having provided supervision on many research works. I wish also to thank Prof. Marco D'Orazio for his precious suggestions on various research aspects, especially regarding ventilated envelopes and users' behavior.

The present works report original material and also new elaborations of data reported on more than 20 papers published by the author in academic field on Elsevier journals and coauthored with: Prof. Costanzo Di Perna, Prof. Placido Munafò, Prof. Marco D'Orazio, Prof. Francesca Tittarelli, Eng. Andrea Ursini Casalena, Eng. Elisa Tomassoni, Eng. Alessio Mastrucci, Eng. Cecilia Bonfigli, Eng. Ambra Vegliò, and Eng. Giacomo Politi.

Thanks to Kenneth P. McCombs, Senior Acquisition Editor Elsevier, for his valuable suggestions that have really improved the quality of the work.

Thanks to Eng. M. Urbinati, deputy director and manager of the Regional Public Housing Authority of Ancona, for funding research studies and for his interest in the research results.

Many thanks also to Centrolegno, Stil casa costruzioni, Alba Costruzioni, and Halfen, all companies working on the building sector that have provided funds and support for the experimental activities.

SYMBOLS, UNITS, AND CONVENTIONS

A	Area	m^2
C	Heat capacity	J/K
R	Thermal resistance	m^2 K/W
T	Period of the variations	s
U	Thermal transmittance under steady state boundary conditions	$W/(m^2\ K)$
Y_{12}	Periodic thermal transmittance	$W/(m^2\ K)$
Z_{ee}	Heat transfer matrix environment to environment	—
Z_{12}	Element of the heat transfer matrix	—
κ_1	Internal areal heat capacity	$kJ/(m^2\ K)$
c	Specific heat capacity	$J/(kg\ K)$
d	Thickness of a layer	m
f	Decrement factor	—
j or i	Unit on the imaginary axis for a complex number; $i = \sqrt{-1}$	—
q	Density of heat flow rate	W/m^2
t	Time	s or h
Δt	Time shift: time lead (if positive), or time lag (if negative)	s or h
λ	Design thermal conductivity	$W/(m\ K)$
ρ	Density	kg/m^3
v	Mean air velocity in the cavity	m/s
H	Height of the cavity	m
ρ_e, ρ_i	External and internal air density	kg/m^3
g	Gravitational acceleration	m/s^2
f_r	Coefficient of friction	—

As a convention within book, based on Fourier's law, the heat flow rate is defined as positive: IN WINTER when it is outgoing, IN SUMMER when it is entering in the room.

CHAPTER ONE

High Thermal Resistance Versus High Thermal Capacity: The Dilemma

1.1 INTRODUCTION

The heating and cooling load of a building is mostly due to the heat transfer across its envelope and thus selecting an appropriate envelope is one of the most effective ways to achieve energy saving. For many years, improving the thermal performance of envelopes meant adopting a thick insulation layer, regardless its position or the presence of mass. These new envelopes act as thermal barriers causing summer overheating and attributing the regulation of indoor comfort to conditioning systems. Forty years of researches have demonstrated that thermal inertia is one of the most important parameters to improve thermal comfort as well as to reduce cooling energy demands especially in particularly dynamic external and internal (e.g., used intermittently) environments. On the other hand, the recent literature on sustainability highlights that, considering economic and environmental aspects, lightweight solutions should be preferred.

This chapter explains these new open issues on the recently introduced superinsulated envelopes also reporting items of previous research in the area. It outlines the book purposes and announces the main innovations and utilities.

1.2 BACKGROUND

1.2.1 The optimal envelope identification is still a challenge

A thermal mass exposed to the external and internal environments responds in a both immediate and time-dependent way, having a certain

Thermal Inertia in Energy Efficient Building Envelopes.
DOI: http://dx.doi.org/10.1016/B978-0-12-813970-7.00001-7

capacity for heat storage. If the temperature remains constant, the mass does not show its dynamic behavior, while for highly variable temperatures it strongly interacts with the environment. The layers relative position and materials adopted influence the time-dependent transmission. Hence, thick insulation layers placed adjacent to the mass inevitably modify this dynamic interaction.

New buildings are subjected to increasingly stringent standards of insulation, regardless the specific climate. The EU regulations on energy saving have been implemented in all Member States with the adoption of the North-European superinsulated model which, focusing on winter heating consumption, has led even in warm countries to the construction of buildings not much related to their climatic context and often designed with disregard for the occupants' needs. Even in such climates, lightweight and superinsulated envelopes have been adopted in new constructions, while in existing buildings retrofit, insulation layers with considerable thicknesses were placed either on the external or internal side of the envelope, regardless of the relative position between mass and thermal insulation. This gave rise to problems of environmental control and to the consequent adoption of expensive systems to reach comfortable conditions in summer.

Indeed, in such hyperinsulated envelopes the opaque walls give small contribution on the thermal heat gains/losses, while the glazed surfaces are responsible for the main quote of internal gains for the greenhouse effect (Fig. 1.1). The glazed surfaces allow the incoming of short wave sun radiation that is then absorbed in the internal room components. Consequently, they reirradiate a long-wave thermal radiation, at which the glass is no more transparent so resulting in a rising of indoor

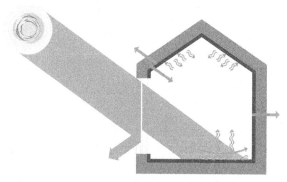

Figure 1.1 Greenhouse effect in superinsulated buildings.

temperature. The recent tendency to realize big transparent surfaces also combined with thermal blocking technologies for glazing (e.g., low-e and solar control smart coatings) and the increasing adoption of new airtight frames techniques, have even more increased the summer overheating risk. Then the heat accumulates inside and the new hyperinsulated envelopes, behaving as thermal barrier, obstacle its dissipation toward the outside.

On the other hand, in the last years the achievement of high levels of thermal comfort has become a priority. Recently, the European Directives 2010/31/EU [1], 2012/27/EU [2], and Standard EN 15251 [3] highlighted the increasing proliferation of air conditioning systems in European countries and stressed the importance to return to an envelope design more strictly linked to the specific climate, also considering the indoor environmental conditions in order to enhance the comfort levels especially in summer.

However, the best solution(s) identification is still an open issue. Many authors have already shown that different insulation—mass configurations have unequal and often opposite effects on the various aspects among energy efficiency [4,5], comfort [6—8], environmental impact, or costs [9,10]. So that the best envelope could be: with internal insulation, in studies for cold climates or only focused on winter performance [11,12]; with internal mass and external insulation, in studies focused on summer performance [13,14—18]; with insulation placed on both sides of the wall [11,15,19,20] or a lightweight solution, in studies on the life cycle and economic assessment [21—26]. Very rarely studies addressed the multidisciplinary simultaneous evaluation of the different aspects.

Other factors complicate the debate on the envelope optimal choice. Firstly, the envelope performance varies based on specific building features, the considered operational conditions among intermittent use and continuous use [11,13,27] and the climate, extreme or with high temperature range [28]. Moreover, in the last decades the assembly techniques and the indoor environmental performance requested by the standards and expected by the occupants, the patterns of occupancy and plant operations have underwent to deep changes, making the identification of the best solution more difficult. These changes are still taking place. Finally, regarding the selection of wall layers, the authors until today [11—14] agreed on the choice of walls that strongly decremented the incoming heat wave thanks to the alternation of capacitive and resistive layers. However, the recent adoption of very thick insulation layers combined with new highly performing materials to reach the requested very high thermal resistances has introduced a new kind of building envelope, with too much elevated attenuating attitude, not achievable with the envelopes of the past. These new solutions have a strongly decoupled behavior between the external and the internal side and behave as thermal barriers, thus blocking not only the incoming but also the outgoing heat flux and creating a "thermos effect," especially during the hot and intermediate seasons.

In all cases, the selection of the optimal strategy is very complex for the strong nonlinearity of the processes involved. This is due to the interaction of dynamic factors, such as the storage effect of massive layers and the strongly variable interior environment strictly linked to the particular (and not easily predictable) behavior of the occupants.

Hence the quantification of consumptions, comfort levels, and environmental impacts of new highly energy efficient envelopes for different ventilation paths, occupants behavior, and timetable for the heating plants also at the varying of the climates is still an open issue.

1.2.2 Comfort issues

The comfort issue in the last years has become a priority. In low-energy buildings, the small range between heating consumptions of the worst and best solution determines that the comfort in unconditioned period and environmental aspects prevail on energy issue [29].

Between these two aspects, the former becomes a priority in temperate climates with hot dry summer for the presence of extensive periods with high temperatures. During such periods, the requirements are different: in the daytime, during the hottest hours, it is necessary to reduce the thermal peaks that are mainly due to the internal loads (for occupants, greenhouse effect from glazed surfaces, etc.), while at night the heat stored during the day should be released toward the outdoor environment.

Therefore it is not advisable to adopt considerable thicknesses of insulating material, which creates a thermos effect impeding an outgoing heat flux. In many cases the use of these envelopes has led to overheating problems not only in temperate climates but also in hottest periods of the cold ones [30—33] and consequently to the need to install expensive cooling and mechanical ventilation systems in order to regulate the indoor environment. In fact while in cold climates, the only requirement is to prevent the heat loss in buildings characterized by an almost stationary behavior throughout the year, in climates characterized by both seasonal and daily high temperature ranges the adopted solutions should be capable of alternatively maximize the thermal barrier effect and the heat loss [34].

The traditional architectures are an example of a very close relationship with the specific climate. Indeed, in dynamic climates they have adapted to the external environment variability (Figs. 1.2 and 1.3) without the use of the systems but through the adoption of passive strategies such as massive walls [35,36] and natural cross-ventilation [37]. Thus in such climates solutions with a deep relation with both the external specific climate and the internal environment should be selected.

As well established in literature, a wall with high thermal capacity on the inner side and the insulation layer on the external side is the best solution since the outer insulation provides the heat loss reduction, while the

Figure 1.2 Traditional buildings in a temperate climate have elongated volumes. The envelopes are massive and the windows are small. *From Energy Build. 88 (February 1, 2015) 367–383, ISSN 0378-7788.*

Figure 1.3 Another example of traditional building in temperate climates. The windows have wooden shutters with adjustable slats and there are shading porches on the southern side.

inner massive layer dynamically interacts with the inside. However, this choice alone could not guarantee appropriate comfort levels since the very thick insulations placed behind the mass impede the outgoing heat flux thus preventing its nighttime cooling and interfering with its dynamic behavior [38,39]. Therefore mixed solutions should be identified which, while using considerable thermal resistance (for the winter), allow an interaction of the massive layers with both the indoor and outdoor environments that varies according to needs.

For new and retrofitted envelopes, various authors demonstrated that dynamic configurations should be preferred. For instance, walls with considerable heat capacity and with thin insulation layers; dynamic walls with seasonal deactivation of the insulation layer; walls with recently developed dynamic finishing phase change materials (PCM); ventilated facades or solar walls used as passive cooling systems.

Between the abovementioned solutions, the slightly insulated walls are not suited for both summer and winter period [6,34]. The dynamic insulation [40−44] is such as a layer of permeable porous insulating material. It is a type of "breathing wall" where the air passes through the porous insulation. It is mainly designed to enhance the indoor ventilation rather than maximize the dynamic behavior of the massive layers. PCM materials [45] are solutions working on the latent heat storage rather than on the maximum exploitation of the traditional massive layers. It could give benefices if choosing an appropriate phase change temperature range.

The ventilated facades may no longer be a solution. In hot-summer Mediterranean climates, ventilated facades (especially those with opaque cladding) have aroused great interest in the last decades because they have resolved the problem of the durability of the outer finishes of the external insulation layer (mainly caused by cracking as a result of aggressive solar radiation) [46]. Ventilated facades with opaque cladding (Fig. 1.4) are generally characterized by the presence of a continuous insulation layer placed adjacent to the internal mass and an external protective cladding fastened to the wall through mechanical anchorages. A naturally ventilated channel is thus created between the insulation layer and the cladding. This solution proved in the past to be excellent for the summer cooling thanks to the "chimney effect," avoiding the overheating drawbacks caused by double skin glazed facades. Nowadays the increasing thickness of the insulation layer interposed between external air gap and internal mass, has strongly reduced the buoyancy-driven ventilation benefices: the insulation works as a thermal barrier between the air gap and

Figure 1.4 Two buildings with ventilated facades, respectively with (A) clay and (B) titanium external cladding. *(B) From Energy Build. 69 (February 2014), 525—534, ISSN 0378-7788.*

the inner mass and prevents its cooling. In literature, no studies addressed this topic.

Another solution to enhance the dynamic interaction with the environment is the ventilated external insulation layer. It consists in an external insulation separated from the massive wall by a channel that can be open or closed through inferior and superior vents. During the summer, the ventilation of the channel deactivates the insulation layer; in winter, the closing of the duct introduces an additional layer of air in steady state [47].

The system was born in Northern Europe with various patents [48—49] but it has been rarely applied for its installation complexity and for the poor winter thermal performance of the air vents, which are generally made of thin aluminum plates. For this reason, our research group (Munafò and Stazi) has studied a preassembled system (Figs. 1.5 and 1.6) with air vents made of insulating material [50]. This type of system could improve the dynamic behavior of the inner mass but only recently some studies were performed in literature on its performance quantification [51].

The system involves the use of two types of panel, "normal panels" (1) and "special panels" (2). Both of them consist of an outer insulating layer spaced from the internal massive wall (or floor) thanks to the use of spacers made by the same insulating material (3) thus creating an air gap. The anchorage of the panels to the massive support occurs as a normal external insulation, i.e., with adhesive and mechanical anchors both placed in correspondence of the spacers. The system is applicable in both walls and roofs. Vents embedded in the inferior and superior "special panels" allow the opening and closing of the ventilation channel. These

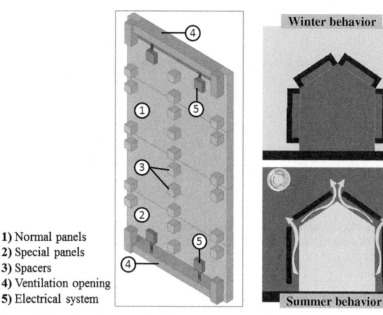

1) Normal panels
2) Special panels
3) Spacers
4) Ventilation opening
5) Electrical system

Winter behavior

Summer behavior

Figure 1.5 Ventilated external insulation layer, patent [50].

Figure 1.6 Prototypes for the ventilated external insulation layer: (A) finishing layers; (B) cubicle spacers within the cavity; (C) special panel with electronic devices; (D) special panel with bottom vertical insulation element against thermal bridges; and (E) special panel with a portion of the metal mesh.

Figure 1.7 (A and B) External views of a solar house in Ancona (central Italy) with ventilated Trombe walls. *From Energy Build. 47 (April 2012) 217−229, ISSN 0378-7788.*

vents (4) are made of insulating material, equipped with seals (the same used for windows) and handled by electronic devices similar to those used for the rolling shutters (5). The system could also include sensors for automatic opening based on the external temperature. The vents opening are provided with expanded metal mesh and insect mesh.

Other dynamic solutions are the passive solar systems, such as the Trombe wall. It is a massive sun capturing system exposed to the southern building side to make full use of the solar energy in winter [52] and that adopts movable dynamic devices (vents, shutters) to enhance the summer cooling (Fig. 1.7). However, the recent introduction of thick insulations on the other building components (walls on the other exposures, roof and ground floor slab) could generate overheating problems [53].

In summary, various authors highlighted the overheating risk of the superinsulated envelopes newly introduced by the energy saving standards, but the quantification of the benefices of restoring the dynamic behavior of the mass, on comfort, consumptions, and global cost is still lacking.

1.2.3 Environmental issues and global costs

Focusing on envelope sustainability, the EU regulations [1,2] also stated that the "measures to improve the energy performance of buildings should take into account cost-effectiveness" and would make possible "to reduce greenhouse gas emissions," introducing economic and environmental issues. To reduce both energy consumption and greenhouse gases emissions related to buildings, it is fundamental to introduce a design approach based on sustainability and to take into consideration every stage of their life. Life cycle assessment (LCA) is a technique used to assess potential environmental impacts of products throughout their life cycle.

A review of the tools for the environmental analysis of buildings can be found in Ref. [54].

In international literature, there are many examples of LCA application on single buildings [55–57], but the comparative LCA of entire buildings characterized by different constructive techniques is very rare [21,22]. Anyhow, there are several studies on the comparative evaluation of single components (walls, load bearing structure, etc.) realized with different constructive systems. Such studies demonstrate that lightweight solutions should be preferred [21–26]. Nevertheless, this type of envelopes, even ensuring a low environmental impact, may not have a positive incidence on energy saving and on other aspects such as comfort or cost effectiveness. Moreover an optimal envelope from energy saving and comfort point of view could result to be disadvantageous from global costs evaluations, for a specific use of the internal environment by the occupants, e.g., for the limited use of the plants [58].

1.3 THE NEED TO RESTORE THE DYNAMIC BEHAVIOR OF THE ENVELOPE

For the abovementioned reasons the integrated study of the impact of a building envelope on different aspects such as energy saving, comfort, environmental sustainability, and cost effectiveness requires a multidisciplinary approach. Numerous authors [24,59–63] addressed this kind of researches but very rarely such studies regarded the simultaneous investigation of all the highlighted aspects.

Moreover for the evaluation of complex phenomena occurring in massive envelopes two basic prerequisites are necessary:

1. Reliable experimental data for the dynamic behavior of the mass also in relation to the complex interaction with highly variable indoor conditions. This requisite is achievable only through experimentations on real case studies and on full-scale test cells.

2. Numerical procedures for the inclusion of these phenomena in the envelope design and for the extension to other situations and climates.

For that reason, the evaluation needs complex transient tools, also calibrated with experimental measures obtained on real case studies.

The aim of this book is to quantify the effect on energy consumptions, comfort levels, and environmental sustainability of the adoption of high insulation thicknesses in both existing and new envelopes. It focuses on the interaction with dynamic strategies such as thermal inertia and natural ventilation. It also addresses the conservation state of the preexisting insulations layers (if any). Moreover for each type of envelope, optimal solutions are drawn at the varying of climate (mild or extreme), building typology (skin dominated or core dominated buildings), mode of use of the internal environment (heating profiles and cross-ventilation).

To achieve these goals, the study adopted a multidisciplinary approach involving integrated phases of measures and simulations:

1. Experimental investigations on real case studies and using an on-site prototype (mock-up), to explore in detail the behavior of different envelopes during the winter and the summer in real occupancy conditions and under controlled conditions. Some surveys on different envelopes were simultaneous. The measured data were also useful to calibrate the simulation models.

2. Laboratory tests on existing insulation (if any), through samples extracted on-site, to quantify the actual performance.

3. Numerical simulations in dynamic regime to extend the boundary conditions to other climates and usage profiles and to consider alternative insulation strategies. The simulations made it possible to quantify comfort levels and energy consumptions extensively for each case study, to define the most beneficial mutual position between mass and insulation in various conditions, and to verify the effect of the introduction of natural ventilated cavities adjacent to the envelope mass.

4. Evaluation of solutions sustainability through the quantification of environmental—economic impacts with LCA analysis; global cost comparison between different scenarios.

5. Integrated evaluations between the various aspects (comfort, energy saving, sustainability).

1.4 WHAT IS NEW...

The book quantifies the differences between various envelope types and demonstrates that in energy efficient solutions the dynamic strategies (thermal mass and ventilation) have low influence on energy saving, while they have a great impact on comfort levels and environmental burdens,

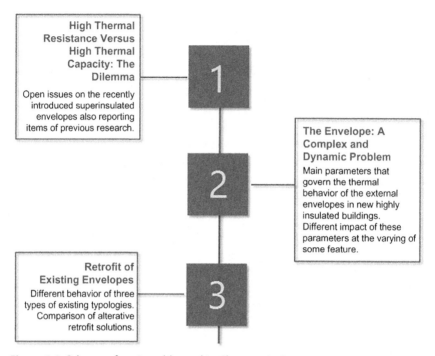

Figure 1.8 Scheme of topics addressed in Chapters 1–3.

with a sometimes-conflicting incidence on these two aspects. Figs. 1.8 and 1.9 report a schematic identification of the topics addressed in each chapter.

Chapter 2, The Envelope: A Complex and Dynamic Problem, introduces the concept of superinsulation and establishes the importance to restore the dynamic behavior of the mass. The text highlights that the importance of thermal capacity varies depending on building geometry, usage typology, relevance of the internal gains, and climates. The main innovation of this chapter is the identification of a particular parameter that takes into account the internal inertia in a simple and effective way: the internal areal heat capacity κ_1 (as defined in EN ISO 13786:2007). Moreover, the identification of an inferior limit for the decrement factor to avoid overheating is novel. The chapter includes a proposal for limit values of the parameter κ_1 combined with the decrement factor f, to ensure comfort and energy saving for highly energy efficient envelopes.

Chapter 3, Retrofit of Existing Envelopes, regards the existing envelopes retrofitting. It highlights that over the years there has been a change in the method in which the traditional envelopes behave toward the

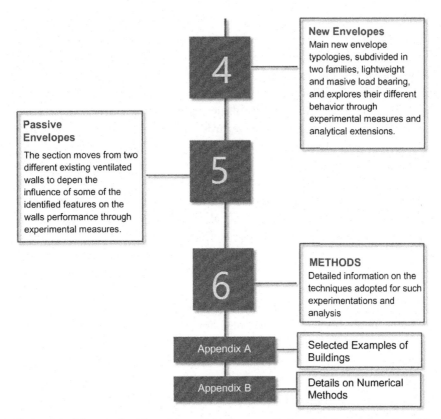

Figure 1.9 Scheme of topics addressed in Chapters 4–6 and in Appendices A and B.

external environment, from "capacitive" (solid brick masonries with high masses) to "stratified" (multilayered cavity walls) and finally to highly thermal "resistant" (introducing thermal barrier insulation layers). The chapter reports simultaneous experimental investigations on three envelopes representative of each type. It also identifies optimal retrofit solutions at the varying of the climate and the mass-insulation materials and positions. Finally, it quantifies the performance of different insulation materials after 20–30 years of service life. The latter aspect is highly innovative since it is based on measured values after several years of service life (through samples extracted on real buildings), while the other studies are in general carried out through laboratory-induced aging.

Chapter 4, New Envelopes, regards the new envelopes solutions. This category includes lightweight timber framed envelopes, cross-laminated timber envelopes, wood-cement walls, and insulated masonry with clay blocks. The chapter reports experimental data, energy consumptions, comfort levels, and environmental impacts. Optimization solutions are suggested (with the relative benefit quantification) for the walls with the worst outcomes. The main innovation of this chapter stands on the broad quantitative data based on experimental surveys, presented on the selected envelopes (that in some cases are still new in the market).

Chapter 5, Passive Envelopes, regards the passive solutions. This category includes ventilated facades, solar walls, and Trombe walls. The chapter reports experimental and numerical data on various seasons to evaluate the incidence of the main design choices: wall exposure respect to the sun and the prevailing wind, chimney height, materials adopted, and presence of external shadings. It also identifies solutions that reduce the environmental impact. A new issue introduced by this chapter is the quantification of the benefices of adopting a passive system at the varying of insulation level on the other building components.

Chapter 6, Experimental Methods, Analytic Explorations, and Model Reliability, reports the methods adopted for each research phase. The method itself embodies the main innovation of the book since all the presented data are based on (and calibrated with) extensive experimental campaigns on real case studies, that are very rare in the existing literature. In particular this chapter:

- reports the general approach and the phases of the multidisciplinary method;
- identifies the 12 case studies commented within the book and identifies their main features (data on general geometry, floors, and vertical walls layers);
- reports the methods adopted for experimental surveys, numerical analyses, environmental evaluations, and global costs. Each method is fully explained reporting the relative standards, to be also reproducible by expert readers.

Appendix A reports detailed information on selected case studies. It deepens the information for all the 12 case studies commented throughout the book (yet briefly presented in Chapter 6: Experimental Methods, Analytic Explorations, and Model Reliability) and introduces other 6 new case studies. For all the case studies, this appendix reports a data sheet with detailed description of envelope dynamic characteristics and the main results of the experimental campaigns.

Appendix B describes in detail the numerical methods, either those adopted for the dynamic calculation of the thermal parameters (according to ISO 13786:2007), or for simulations with Energy Plus software.

The performance quantification of different types of envelope in various situations can provide the building designer with a better prediction of in-service comfort levels, energy saving, and environmental impact. But perhaps of greater importance, it provides a clear picture of the interaction between envelope and internal/external environment and a better understanding of why new and existing envelopes behave as they do. The techniques described and illustrated in this book may be readily extended to include a wide range of additional envelopes.

REFERENCES

[1] Directive 2010/31/EU of the European Parliament and of the Council of 19 May 2010 on the Energy Performance of Buildings. Official Journal of the European Union.

[2] Directive 2012/27/EU of the European Parliament and of the Council of 25 October 2012 on Energy Efficiency, Amending Directives 2009/125/EC and 2010/30/EU and Repealing Directives 2004/8/EC and 2006/32/EC. Official Journal of the European Union.

[3] European Standard EN 15251, Indoor Environmental Input Parameters for Design and Assessment of Energy Performance of Buildings Addressing Indoor Air Quality, Thermal Environment, Lighting and Acoustics, 2007.

[4] S. Ferrari, Building envelope and heat capacity: re-discovering the thermal mass for winter energy saving, in: Conference Proceedings: "Building Low Energy Cooling and Advanced Ventilation Technologies in the 21st Century", Crete island, Greece, 2007, pp. 346−351.

[5] N. Aste, A. Angelotti, M. Buzzetti, The influence of external walls thermal inertia on the Energy performance of well insulated buildings, Energy Build. 41 (2009) 1181−1187.

[6] D.M. Ogoli, Predicting indoor temperatures in closed buildings with high thermal mass, Energy Build. 35 (2003) 851−862.

[7] C.A. Balaras, The role of thermal mass on the cooling load of buildings. An overview of computational methods, Energy Build. 24 (1996) 1−10.

[8] V. Cheng, E. Ng, B. Givoni, Effect of envelope color and thermal mass on indoor temperatures in hot humid climate, Solar Energy 78 (2005) 528−534.

[9] S.A. Al-Sanea, M.F. Zedan, Improving thermal performance of building walls by optimizing insulation layer distribution and thickness for same thermal mass, Appl. Energy 88 (2011) 3113−3124.

[10] G. Kumbaroğlu, R. Madlener, Evaluation of economically optimal retrofit investment options for energy savings in buildings, Energy Build. 49 (2012) 327−334.

[11] M. Bojić Lj, D.L. Loveday, The influence on building thermal behaviour of the insulation/masonry distribution in a three-layered construction, Energy Build. 26 (1997) 153−157.

[12] P.T. Tsilingiris, Wall heat loss from intermittently conditioned spaces − the dynamic influence of structural and operational parameters, Energy Build. 38 (2006) 1022−1031.

[13] E. Kossecka, J. Kosny, Influence of insulation configuration on heating and cooling loads in a continuously used building, Energy Build. 34 (2002) 321−331.

[14] P.T. Tsilingiris, Parametric space distribution effects of wall heat capacity and thermal resistance on the dynamic thermal behavior of walls and structures, Energy Build. 38 (2006) 1200−1211.

[15] S.A. Al-Sanea, M.F. Zedan, Effect of thermal mass on performance of insulated building walls and the concept of energy savings potential, Appl. Energy 89 (2012) 430−442.

[16] K. Gregory, B. Moghtaderi, H. Sugo, A. Page, Effect of thermal mass on the thermal performance of various Australian residential constructions systems, Energy Build. 40 (2008) 459−465.

[17] J. Zhou, G. Zhang, Y. Lin, Y. Li, Coupling of thermal mass and natural ventilation in buildings, Energy Build. 40 (2008) 979−986.

[18] F. Stazi, C. Di Perna, P. Munafò, Durability of 20-year-old external insulation and assessment of various types of retrofitting to meet new energy regulations, Energy Build. 41 (2009) 721−731.

[19] H. Asan, Effects of wall's insulation thickness and position on time lag and decrement factor, Energy Build. 28 (1998) 299−305.

[20] K.J. Kontoleon, D.K. Bikas, The effect of south wall's outdoor absorption coefficient on time lag, decrement factor and temperature variations, Energy Build. 39 (2007) 1011−1018.

[21] G. Pajchrowski, A. Noskowiak, A. Lewandowska, W. Strykowski, Materials composition or energy characteristic. What is more important in environmental life cycle of buildings? Build. Environ. 72 (2014) 15−27.

[22] S. John, B. Nebel, N. Perez, A. Buchanan, Environmental Impacts of Multi-Storey Buildings Using Different Construction Materials, New Zealand Ministry of Agriculture and Forestry, 2009.

[23] H. Monteiro, F. Freire, Life-cycle assessment of a house with alternative exterior walls: comparison of three impact assessment methods, Energy Build. 47 (2012) 572−583.

[24] A. Dodoo, L. Gustavsson, R. Sathre, Effect of thermal mass on life cycle primary energy balances of a concrete and a wood-frame building, Appl. Energy 92 (2012) 462−472.

[25] L. Guardigli, F. Monari, M.A. Bragadin, Assessing environmental impact of green buildings through LCA methods: a comparison between reinforced concrete and wood structures in the European context, Proc. Eng. 21 (2011) 1199−1206.

[26] R. Broun, G.F. Menzies, Life cycle energy and environmental analysis of partition wall systems in the UK, Proc. Eng. 21 (2011) 864−873.

[27] K. Ulgen, Experimental and theoretical investigation of effects of wall's thermos-physical properties on time lag and decrement factor, Energy Build. 34 (2002) 273−278.

[28] Y. Huang, J. Niu, T. Chung, Study on performance of energy-efficient retrofitting measures on commercial building external walls in cooling-dominant cities, Appl. Energy 103 (2013) 97−108.

[29] F. Stazi, E. Tomassoni, C. Bonfigli, C. Di Perna, Energy, comfort and environmental assessment of different building envelope techniques in a Mediterranean climate with a hot dry summer, Appl. Energy 134 (2014) 176−196.

[30] R.S. McLeod, C.J. Hopfe, A. Kwan, An investigation into future performance and overheating risks in Passivhaus dwellings, Build. Environ. 70 (2013) 189−209.

[31] R. Lindberg, A. Binamu, M. Teikari, Five-year data of measured weather, energy consumption, and time-dependent temperature variations within different exterior wall structures, Energy Build. 36 (2004) 495−501.

[32] A. Norén, J. Akander, E. Isfält, The effect of Thermal Inertia on Energy Requirement in a Swedish Building − Results Obtained with Three Calculation Models, Low Energy Sust. Build. 1 (1999) 1−16.

[33] C. Di Perna, F. Stazi, A. Ursini Casalena, M. D'Orazio, Influence of the internal inertia of the building envelope on summertime comfort in buildings with high internal heat loads, Energy Build. 43 (2011) 200−206.

[34] Y. Zhang, Y. Zhang, X. Wang, Q. Chen, Ideal thermal conductivity of a passive wall: determination method and understanding, Appl. Energy 112 (2013) 967–974.

[35] N. Cardinale, G. Rospi, A. Stazi, Energy and microclimatic performance of restored hypogenous buildings in south Italy: the "Sassi" district of Matera, Build. Environ. 43 (2010) 94–106.

[36] Z. Yilmaz, Evaluation of energy efficient design strategies for different climatic zones: comparison of thermal performance of buildings in temperate-humid and hot-dry climate, Energy Build. 39 (2007) 306–316.

[37] A. Gagliano, F. Patania, F. Nocera, C. Signorello, Assessment of the dynamic thermal performance of massive buildings, Energy Build. 72 (2014) 361–370.

[38] F. Stazi, C. Bonfigli, E. Tomassoni, C. Di Perna, P. Munafò, The effect of high thermal insulation on high thermal mass: is the dynamic behaviour of traditional envelopes in Mediterranean climates still possible? Energy Build. 88 (2015) 367–383.

[39] K.M.S. Chvatala, H. Corvachob, The impact of increasing the building envelope insulation upon the risk of overheating in summer and an increased energy consumption, J. Build. Perform. Simul. 2 (4) (2011) 267–282.

[40] F. Stazi, A. Vegliò, C. Di Perna, P. Munafò, Experimental comparison between 3 different traditional wall constructions and dynamic simulations to identify optimal thermal insulation strategies, Energy Build. 60 (2013) 429–441.

[41] M.S.E. Imbabi, A passive–active dynamic insulation system for all climates, Int. J. Sustain. Built Environ. 1 (2012) 1247–1258.

[42] B.J. Taylor, R. Webster, M.S. Imbabi, The building envelope as an air filter, Build. Environ. 23 (1999) 353–361.

[43] G. Gan, Numerical evaluation of thermal comfort in rooms with dynamic insulation, Build. Environ. 35 (2000) 445–453.

[44] A. Dimoudi, A. Androutsopoulos, S. Lykoudis, Experimental work on a linked, dynamic and ventilated, wall component, Energy Build. 36 (2004) 443–453.

[45] D. Zhou, C.Y. Zhao, Y. Tian, Review on thermal energy storage with phase change materials (PCMs) in building applications, Appl. Energy 92 (2012) 593–605.

[46] M. Ciampi, F. Leccese, G. Tuoni, Ventilated facades energy performance in summer cooling of buildings, Solar Energy 75 (2003) 491–502.

[47] H. Bartodziej, Method of Heat Flow Control Through an External Wall of Building and Wall Assembly for Execution of This Method, WO00/60183A1, (Patent), 2000.

[48] G. Anmelder, Fassadenwärme-Dämm-Verbundsystem, DE102004001601A1, (Patent), 2005.

[49] R. Güldenpfenning, Verfahren Herstellung wärmegedämmter Putzfassaden, DE3238445A1, (Patent), 1984.

[50] P. Munafò, F. Stazi, Patent n. MI2011A001317, concession n. 001407018/2014, 2011.

[51] F. Stazi, A. Vegliò, C. Di Perna, P. Munafò, Retrofitting using a dynamic envelope to ensure thermal comfort, energy savings and low environmental impact in Mediterranean climates, Energy Build. 54 (2012) 350–362.

[52] O. Saadatian, K. Sopian, C.H. Lim, N. Asim, M.Y. Sulaiman, Trombe walls: a review of opportunities and challenges in research and development, Renewable Sustainable Energy Rev. 16 (2012) 6340–6351.

[53] F. Stazi, A. Mastrucci, C. di Perna, The behaviour of solar walls in residential buildings with different insulation levels: an experimental and numerical study, Energy Build. 47 (2012) 217–229.

[54] I.Z. Bribián, A.A. Usón, S. Scarpellini, Life cycle assessment in buildings: State-of-the-art and simplified LCA methodology as a complement for building certification, Build. Environ. 44 (2009) 2510–2520.

[55] S. Nibel, T. Luetzkendorf, M. Knapen, C. Boonstra, S. Moffat, Annex 31: Energy Related Environmental Impact of Buildings, Technical Synthesis Report, International Energy Agency.

[56] M. Erlandsson, M. Borg, Generic LCA—methodology applicable for buildings, constructions and operation services—today practice and development needs, Build. Environ. 38 (no. 7) (2003) 919−938.

[57] A. Haapio, P. Viitaniemi, A critical review of building environmental assessment tools, Environ. Impact Assess. Rev. 28 (2008) 469−482.

[58] A.L. Pisello, F. Asdrubali, Human-based energy retrofits in residential buildings: a cost-effective alternative to traditional physical strategies, Appl. Energy 133 (2014) 224−235.

[59] S. Thiers, B. Peuportier, Energy and environmental assessment of two high energy performance residential buildings, Build. Environ. 51 (2012) 276−284.

[60] P. Mendonca, L. Braganca, Sustainable housing with mixedweight strategy—a case study, Build. Environ. 42 (2007) 3432−3443.

[61] R.M. Pulselli, E. Simoncini, N. Marchettini, Energy and emergy based cost-benefit evaluation of building envelopes relative to geographical location and climate, Build. Environ. 44 (2009) 920−928.

[62] S. Chiraratananon, V.D. Hien, Thermal performance and cost effectiveness of massive walls under thai climate, Energy Build. 43 (2011) 1655−1662.

[63] R. Morrissey, E. Horne, Life cycle cost implications of energy efficiency measures in new residential buildings, Energy Build. 43 (2011) 915−924.

CHAPTER TWO

The Envelope: A Complex and Dynamic Problem

2.1 INTRODUCTION

This chapter introduces the main parameters that govern the thermal behavior of the external envelopes in new highly insulated buildings. In particular it identifies the internal areal heat capacity κ_1 and decrement factor f as the main influencer of the summer behavior, while the steady state thermal transmittance U and the decrement factor f of the winter performance. Combined limit values for κ_1 and f are proposed. Moreover the different impacts of these parameters at the varying of some feature, such as the exposed envelope area, the presence of shading/overhangs, the user behavior regarding the heating plants operation profile, the natural cross-ventilation, and the climate zone are outlined. The chapter includes experimental measures on a vernacular building with different mass at the two levels and ends with an experimental evidence of the importance of the occupants' behavior.

2.2 RELEVANT PARAMETERS

2.2.1 Problem description

The influence of the external walls layout not only on energy saving but also on comfort levels was first observed about 40 years ago, and research to deepen the most suitable thermophysical properties has increased at an increasing rate ever since. The volume of literature on this subject is so vast and it is not possible here to provide a detailed review. For an historic overview and a detailed description of the causes and effects of the envelope changes from climate-modifying membrane as in traditional buildings to the elimination of geographical differences in actual

Thermal Inertia in Energy Efficient Building Envelopes.
DOI: http://dx.doi.org/10.1016/B978-0-12-813970-7.00002-9

"international" building constructive typologies, the books by Burberry [1] and Bansal [2] are recommended.

The present section demonstrates that the dynamic behavior of the mass could be guaranteed with a careful selection of the thermal parameters. Solutions with high internal heat capacity and a remarkable thermal damping attitude (but not the highest achievable one) are preferable in both winter and summer for different properties of the mass.

Consider the walls shown in Table 2.1. All of them present very low stationary thermal transmittance values, $U = 0.24-0.26$ W/(m^2 K). Thermal transmittance is a measure of heat loss through a building component. The very low U value adopted determines that the outgoing and incoming fluxes through the envelope are strongly reduced and therefore the internal conditions are much less bonded to the external ones. Thus the internal gains (for greenhouse effect through the glazed surfaces, occupants, equipment, etc.) assume a great incidence.

Differently in the walls the dynamic parameters were varied:

- For walls W1, W2, and W3 a quite elevated f value was set ($f = 0.13$) so the wall is not very effective at suppressing the temperature swing. The internal areal heat capacity κ_1 was varied from 5 to 40−70 kJ/(m^2 K), leading to considerable differences in the internal inertia of the walls.

- For walls W4 and W5 the internal inertia κ_1 was fixed at a high value (40 kJ/m^2 K) and two different levels for decrement factor f were adopted, respectively 0.038 and 0.027 (the latter is a very low value, indicating elevated damping effect); the wall W6 adopts a low κ_1, 20 kJ/(m^{2}K), and a high f value (0.163) to explore the case expected to be the worst choice.

2.2.2 Dynamic thermal characteristics of the building envelope

The evaluation of the dynamic thermal performance of the building envelopes has been done according to EN ISO 13786:2007 [3]. The detailed method description is reported in Appendix B.

At the cross section of the external envelope the temperature profiles are variable for each instant, depending on inside temperatures, outside temperature, and thermophysical properties of the layers. The dynamic thermal characteristics, reported in Table 2.1, are obtained by imposing a conductive thermal exchange condition and having for each material the values of the thermal conductivity, specific heat, density, and thickness.

Table 2.1 Comparison of walls with different dynamic properties

	W1	W2	W3	W4	W5	W6
Type of mass	Masonry	Masonry	Masonry	Masonry	Wood-cement	Wood
Wall thickness (cm)	43	40	25	49	43	28
U (W/(m^2 K))	0.24	0.24	0.24	0.26	0.26	0.26
Y_{12} (W/(m^2 K))	0.03	0.03	0.03	0.01	0.007	0.043
κ_1 (kJ/m^2 K)	5	40	70	40	40	20
f (−)	0.13	0.13	0.13	0.038	0.027	0.163
Insulation						**External/internal**
Thickness (cm)	13.1	20.2	10.2	10	—	8/5
Conductivity (W/(m K))	0.036	0.06	0.027	0.048	—	0.052/0.042
Density (kg/m^3)	35	134	35	230	—	200/40
Specific heat (J/(kg K))	1200	1800	1200	2100	—	2100/670
Mass						
Thickness (cm)	30.0	20.0	15.0	35.0	38	8.5
Conductivity (W/(m K))	0.8	0.3	0.7	0.24	0.107	0.12
Density (kg/m^3)	1575	850	1700	800	887	500
Specific heat (J/(kg K))	1000	840	1820	840	1282	1600

This method allows relating heat flow rate on one side (e.g., internal lining) to temperature variation on the same side (due to internal heat gains) and also computing the dynamic transfer properties through the whole component relating the physical quantities on one side of the component to those on the other side. These two calculations, combined together give the heat capacity of a component and thus its heat storage ability.

The periodic analysis method hypothesizes a sinusoidal temperature variation on one side of the envelope keeping the other side temperature constant. To evaluate the envelope response with regard to a daily temperature variation, a sinusoidal period equivalent to 24 hours has been considered. The effect of this imposed condition is reflected on the adjacent layers through a heat flow rate, assumed to be one dimensional and with sinusoidal trend too. During this dynamic interaction the sinusoidal heat wave flows from one side to the other. The heat transfer is function of the periodic penetration depth δ of each layer, thus depending on thermal diffusivity of the material a, and the rate of change in material temperature θ with respect to time t (Table 2.2. points (1)−(3)).

A generic sinusoidal function in which amplitude, angular frequency (ω), and initial phase are time-invariant could be studied through complex numbers. To deepen this aspect see the note on complex number in Appendix B. Complex amplitudes are denoted with circumflexes.

The heat transfer matrix Z is used to relate the complex amplitude of temperature and heat flow rate at the one side of a component (e.g., the external side 2) to the complex amplitude of temperature and heat flow rate at the other (e.g., internal side 1).

The concept could be described through the following relation:

$$\begin{pmatrix} \hat{\theta}_2 \\ \hat{q}_2 \end{pmatrix} = \begin{pmatrix} Z_{11} & Z_{12} \\ Z_{21} & Z_{22} \end{pmatrix} \cdot \begin{pmatrix} \hat{\theta}_1 \\ \hat{q}_1 \end{pmatrix}$$

To obtain each element of the heat transfer matrix Z, the calculation of the periodic penetration depth δ and ξ (ratio of the thickness of the layer and δ) are required, according to relations (1) and (3) in Table 2.2.

For a multilayered wall the heat transfer matrix Z is obtained from the product of several individual heat transfer matrices (one for each layer) in

Table 2.2 Dynamic properties according to EN ISO 13786:2008

$$\delta = \sqrt{\frac{\lambda T}{\pi \rho c}} \tag{1}$$

Periodic penetration depth, of a heat wave in a material (m). Where:
T = period of the variation (s)
ρ = density (kg/m^3)
c = specific heat [J/(kg K)]
λ = thermal conductivity [W/(m K)]

$$a = \frac{\lambda}{\rho c} \tag{2}$$

Thermal diffusivity (m^2/s): describes how quickly a material reacts to a change in temperature

$$\xi = \frac{d}{\delta} \tag{3}$$

Ratio of the thickness of the layer to the penetration depth (−). Where:
d = layer thickness (m)

$$Z_{11} = Z_{22} = \cosh(\xi)\cos(\xi) + j\sinh(\xi)\sin(\xi) \tag{4}$$

Complex numbers; elements of heat transfer matrix.

$$Z_{12} = -\frac{\delta}{2\lambda}\{\sinh(\xi)\cos(\xi) + \cosh(\xi)\sin(\xi) \tag{5}$$

$$+ j[\cosh(\xi)\sin(\xi) - \sinh(\xi)\cos(\xi)]\}$$

It is a complex number; its modulus represents the amplitude of the temperature on side 2 when side 1 is subjected to a periodically varying density of heat flow rate with an amplitude of 1 W/m^2

$$Z_{21} = -\frac{\lambda}{\delta}\{\sinh(\xi)\cos(\xi) - \cosh(\xi)\sin(\xi) \tag{6}$$

$$+ j[\sinh(\xi)\cos(\xi) + \cosh(\xi)\sin(\xi)]\}$$

It is a complex number; its modulus represents the amplitude of the density of heat flow rate through side 2 resulting from a periodic variation of temperature on side 1 with an amplitude of 1 K

$$\kappa_1 = \frac{T}{2\pi}\left|\frac{Z_{11} - 1}{Z_{12}}\right| \tag{7}$$

$$\kappa_1 = \frac{1}{\omega}|Y_{11} - Y_{12}| \tag{8}$$

(*Continued*)

Table 2.2 (Continued)

Internal areal heat capacity [kJ/(m² K)]: heat capacity of the internal side divided by area of element. Where:

ω = angular frequency; $\omega = \dfrac{2\pi}{T}$ (rad/s)

$$Y_{12} = -\frac{\hat{q}_1}{\hat{\theta}_2} \tag{9}$$

$$Y_{12} = -\frac{1}{Z_{12}} \tag{10}$$

Periodic thermal transmittance (W/m² K): complex quantity defined as the complex amplitude of the density of heat flow rate through the surface of the component adjacent to the internal side $\hat{q}_1$ (side 1), divided by the complex amplitude of the temperature in the external side of the component $\hat{\theta}_2$ (side 2) when the internal temperature (side 1) is held constant

$$Y_{11} = \frac{\hat{q}_1}{\hat{\theta}_1} \tag{11}$$

$$Y_{11} = -\frac{Z_{11}}{Z_{12}} \tag{12}$$

Thermal admittance (W/m² K): is the amplitude of the density of heat flow on one side $\hat{q}_1$ resulting from a unit temperature amplitude on the same side $\hat{\theta}_1$, when the temperature amplitude on the other side is zero ($\hat{\theta}_2 = 0$)

$$f = \frac{|Y_{12}|}{U_0} \tag{13}$$

Decrement factor (−): ratio of the modulus of the periodic thermal transmittance Y_{12} to the steady-state thermal transmittance U_0 (calculated according to ISO 6946 ignoring any thermal bridges). This is the ratio of the peak heat flow out of the external surface of the element per unit degree of external temperature swing to the steady state heat flow through the element per unit degree of temperature difference. Where: U_0 = stationary thermal transmittance (W/m² K)

$$\Delta t_f = \frac{T}{2\pi} arg(Z_{12}) \tag{14}$$

Time shift (h): time the heat wave takes to propagate from external to internal side. Time lead (if positive) or time lag (if negative).

See Appendix B for the detailed method description. The easy calculation of these parameters could be done through the spreadsheet developed by A. Ursini Casalena freely available at www.mygreenbuildings.org [4].

the correct order, including convection and radiation condition related to the surface resistances of the boundary layers.

The *periodic thermal transmittance* Y_{12} is the cyclic heat flux released on the internal surface of the wall per unit cyclic of temperature variation imposed on its external side while holding a constant indoor temperature (point (9) in Table 2.2). The negative sign is because the Standard 13786 assumes as convention that the heat flow rate is positive when entering the surface of the component. This cyclic flux can be associated to an amplitude and a phase. So two parameters can be related to it: the decrement factor and the time shift.

The *decrement factor f*, relates the amplitude of the cyclic external temperature swing acting on the wall to the periodic heat flux released to the indoor air. It can be seen as the decrease of the temperature amplitude recorded at the internal side respect to its initial amplitude in the outer side. Thus it provides information about the dampening of the periodic thermal signal passing from outside to inside. It is also described as the amplitude of periodic thermal transmittance normalized with respect to steady state thermal transmittance.

The *time shift* is the period of time between the maximum amplitude of a cause and the maximum amplitude of its effect. It could be a time lead (if positive), or a time lag (if negative). It is related to the periodic heat flux phase and gives the delay between a peak in the outdoor temperature profile and the corresponding peak in the heat flux released to the indoor air. It is the phase of the complex number Z_{12}, measured in hours and referred to a solicitation having a period T. The decrement factor is always less than 1. For envelopes characterized by very low global inertia, the periodic thermal transmittance almost equals the steady state value and f tends to unity, while Δt_f tends to 0.

2.2.3 The internal areal heat capacity

The parameter that quantifies the internal inertia of an opaque envelope under dynamic conditions is the *internal areal heat capacity* κ_1. It represents the ability of a building component to store energy from one side (in this case 1 stands for internal side) when the corresponding temperature (inside the room) varies periodically. This parameter is the net periodic thermal conductance divided by angular frequency. It can be seen as the amplitude of density of heat flow rate on the internal side resulting from a unit indoor temperature amplitude, when the external temperature is held constant but excluding the quote of heat flow rate on the same internal side resulting from a unit outdoor temperature amplitude at constant indoor temperature. A higher κ_1 values means a better ability to store heat on the inner side on a given thermal variation. It represents the amount of energy stored in the element over the first half period of the heat flow swing per unit area of element per unit degree of temperature swing. The same amount of heat is released in the following half period.

The graph in Fig. 2.1 reports the internal areal heat capacity of a fictitious material obtained in a wooden envelope (walls of the mock-up reported in Appendix A, case study A1) by adopting as internal lining a double dry clay panel: $\kappa_1 = 33$ kJ/(m^2 K), thickness 44 mm, thermal conductivity $\lambda = 0.47$ W/(m K), specific heat capacity $c = 1000$ J/(kg K), density $\rho = 1300$ kg/m^3. Starting from this material the characteristics regarding thickness, thermal conductivity, specific heat, and density were varied one by one.

The graph shows that κ_1 increases at the increasing of all the parameters (thickness, λ, c, ρ). However, the increase of λ has an almost negligible influence on κ_1 increase. The f value instead has an inverse trend, with decreasing values at the increment of the studied parameters, except for λ.

So a material with high thickness, high specific heat capacity, high density, and low thermal conductivity could simultaneously achieve high periodic heat capacity and low decrement factors (and so a pronounced attenuating attitude).

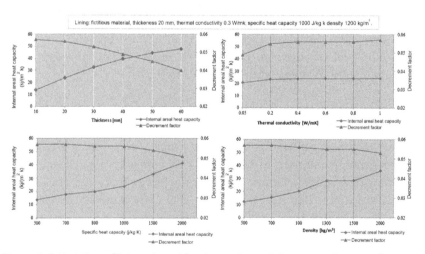

Figure 2.1 Variation of internal areal heat capacity and decrement factor at the varying of the thermal parameter of a fictitious material (thickness, thermal conductivity, specific heat capacity, and density).

The maximization of the internal areal heat capacity is very important during the summer and in the intermediate seasons. In these periods the low outdoor—indoor temperature gradient combined with the presence of a high insulation thickness determines that both external and internal surfaces will tend toward average temperatures thus reducing the heat flux that does not entirely cross the wall. Hence the greatest amount of heat gains comes from the internal side (mainly due to the presence of occupants, equipment, lightings, and solar gain for greenhouse effect). In this case the inner layer adjacent to the room assumes a greater importance than the other layers behind it to enhance the dynamic interaction with the internal overheated environment. This internal layer should be realized with materials able to maximize the heat transfer rate (high λ) and to store it during the hottest hours (high κ_1) releasing it on the internal side during the night. It allows to keep operating and indoor air temperatures lower and to reduce the cooling system working. The higher the thermal barrier effect realized by the envelope outer layers (that creates an overheated environment) or the higher the internal heat gains (high occupancy levels, solar radiation coming from the windows and heat crossing through the opaque wall) the greater the importance of the adoption of high κ_1 values.

2.2.4 Effect of the internal heat capacity on summer performance

The present section demonstrates that an increasing of κ_1 value when all the other parameters are fixed determines a flattening of the surface and operative temperature curves. The surface temperatures of the three walls W1, W2, and W3 (with increasing κ_1 value) were compared for the months of May and September, hypothesizing a naturally vented environment (graphs on Figs. 2.2 and 2.3). The simulations are relative to a school building characterized by high internal heat gains (see Appendix A, case study B7).

During the month of May a wall with low internal inertia (W1) leads to the greatest variations and the highest surface temperatures. This type of solution is not to be favored from the comfort point of view. At the end of the hot period, namely in the first days of September, the difference between the studied walls is even more evident. The solution W3, with the highest internal mass, is to be favored from the point of view of comfort for its lowest temperatures and fluctuations. By comparing the two graphs it can be noticed that in the cooler periods (the first week of May and the last week of September) the three curves intersect with the highest temperatures recorded by W3 and with fluctuations that lower

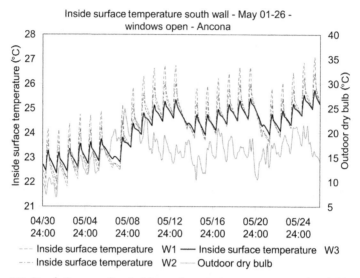

Figure 2.2 Simulation results. Inside surface temperatures, month of May. *From Influence of the internal inertia of the building envelope on summertime comfort in buildings with high internal heat loads, Energy Build. 43 (1) (January 2011) 200—206, ISSN 0378-7788.*

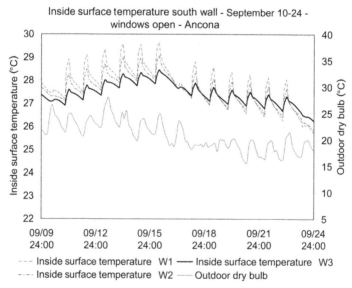

Figure 2.3 Simulation results. Inside surface temperatures, month of September. *From Influence of the internal inertia of the building envelope on summertime comfort in buildings with high internal heat loads, Energy Build. 43 (1) (January 2011) 200—206, ISSN 0378-7789.*

going from W1 to W3. In the warmer periods (end of May and beginning of September) the three curves do not intersect and the line for wall W3 is below the others, with temperature values that are constantly the lowest. The inversion of trend between the three techniques is clearly visible at the first days of May and around 15th of September. So the technique W3 has favorable conditions for both cold periods (avoiding too low temperature values at night) and hot periods (guaranteeing lower values for the whole day).

2.2.5 The dampening attitude

The second important feature for summer performance is the thermal decrement factor, f. It represents the reduction of periodic temperature oscillations amplitude across the fabric as a result of heat storage within that fabric, and hence the rate of heat transfer is attenuated by the factor f. It embodies the envelope capacity to dampen the thermal wave under dynamic conditions (Fig. 2.4).

Low value of the decrement factor means that in summer only a small fraction of the heat wave induced by absorbed solar radiation on the external side will be transmitted to the internal space through the opaque envelope. In winter there will be a dampening effect on the outgoing heat wave.

The higher the thermal capacity or thermal resistance of a material, the stronger is the dampening effect on the crossing heat wave. So the decrement factor is influenced by the thermal diffusivity of each layer and also depends on the layers relative position. For example, some authors demonstrated that the lowest f value is obtained by alternating resistive and capacitive layers [5]; the same findings were outlined by other researchers that demonstrated that to achieve better thermal performance of the wall different wall layers of materials having unequal thermophysical properties should be combined thus obtaining composite walls [6].

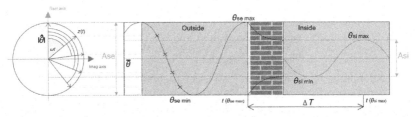

Figure 2.4 Meaning of the decrement factor and time lag. On the left is also reported a projection on the complex plane through phasors (see Appendix B for explanation).

This is because the heat gained by the wall is stored inside the various layers of the structure and trapped thanks to the adjacent highly insulating layers. Then it is dissipated to the inside.

To give a physical meaning to the decrement factor, it is usually expressed as the ratio of the periodic thermal transmittance to the steady state U value. A unitary value of the decrement factor implies that a construction has no attenuation influence on the heat flowing through it, as the cyclic thermal transmittance equals the steady state U value. Lowest decrement factors are therefore favored.

Thus the minimization of this parameter is very important during both the summer and the winter, but with some limits since too low decrement factor values could be inappropriate. The following section deepens this new aspect.

2.2.6 Effect of the dampening attitude on summer performance

The annual comfort analysis carried out on a typical apartment of a residential case study (building project in Fig. 2.10) focusing on operative temperatures, made it possible to identify different phases and a trend inversion for the technologies W4−W6 (as highlighted for wall W1−W3 in the section 2.2.4). The results are reported in Fig. 2.5.

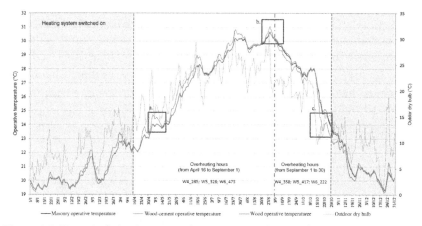

Figure 2.5 Numerical simulations in the temperate climate of Ancona (central Italy). Yearly performance of envelopes with different combinations of thermal dynamic parameters. Masonry corresponds to wall W4, wood-cement to W5, and wood to W6. *From Energy, comfort and environmental assessment of different building envelope techniques in a Mediterranean climate with a hot dry summer, Appl. Energy 134 (December 1, 2014) 176−196, ISSN 0306-2619.*

During the winter (*red hatched area* (gray in print versions)) there is a constant trend with very similar values among the three techniques. In early spring there is a sudden growth as the external temperatures and the diurnal temperature variation increase, with the lowest values for the two massive solutions W4 and W5.

In late spring and in hottest period (until September 1) the graph continues to show a growing trend. The high-mass techniques (mainly the masonry W4 with its higher internal inertia) maintain lower operative temperatures than the wooden solution and consequently better comfort conditions (Figs. 2.6A and B).

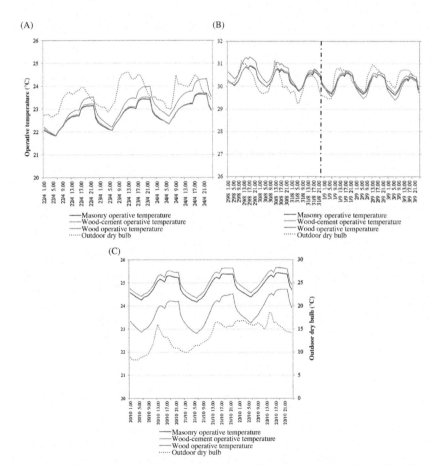

Figure 2.6 Numerical simulations. Enlargement of selected periods from Fig. 2.5 to highlights an inversion of the trend between the three techniques. (A) Spring, (B) summer, and (C) autumn. *From Energy, comfort and environmental assessment of different building envelope techniques in a Mediterranean climate with a hot dry summer, Appl. Energy 134 (December 1, 2014) 176–196, ISSN 0306-2619.*

After this period there is an inversion of the trend (September 1) among the techniques. The wood W6 temperature drops and becomes lower than those of the masonry and wood-cement (Fig. 2.6C) cooling itself more rapidly respect to the other solutions that have a greater seasonal inertia. At the end of warm season until the beginning of the heated period the massive envelopes still record higher internal temperatures than the lightweight technique W6. This is a more favorable behavior for the approaching of the cold period.

In the hottest summer period (until September 1) the sum of overheating hours (reported within Fig. 2.5) is very high for W6, intermediate for W5, and it records the lowest values for W4.

The cooling demand in a temperate climate (city of Ancona) with intermittent operation (2 sections: 5:00 a.m.−9:00 a.m., 3:00 p.m.−11:00 p.m.) is shown in Table 2.3. The data outline a difference of 2.5% in summer between the best and the worst solution, namely W4 and W6. The masonry W4 gave the best results for the summer for its optimum configuration, with high value of internal areal heat capacity (and as a consequence the highest inner inertia) and a low f value. The wood-cement (W5) is worse than the masonry for its too low f value since completely blocking the crossing wave it creates a thermos effect. The wood (W6) is the worst solution for all year round, even if with small differences respect to the other two solutions, since it has the worst dynamic thermal parameters: the lowest internal areal heat capacity (and as a consequence the lowest inner inertia) and the highest f value.

A high internal inertia is also recommended for its moisture buffering effect [7,8]. The results regarding average summer values for the mean indoor relative humidity (RH) for the three techniques show that there are small differences between indoor and outdoor humidity values

Table 2.3 Cooling consumption and indoor relative humidity for envelopes W4, W5, and W6

	Useful energy demand for cooling (kWh prim./m² year)	Average seasonal RH value (summer) (%)
W4: masonry	5.86	72.9
W5: wood-cement	5.93	72.8
W6: wood	6.03	72.7

Summer: CDD = 742; T_{med} = 23.3°C, T_{max} = 36.9°C; Δ°C = 11.7°C. *Ancona* (central Italy, type *Csa*: hot-summer Mediterranean climate) No shading devices.

indicating (as for the heat transmission) the greatest importance of the hygrometric inertia of the inner layer rather than of the entire wall configuration. This also in consideration of the high insulation thicknesses that strongly reduces the vapor crossing. The similar indoor humidity levels recorded for the three techniques depend on the similar hygrometric properties of the internal plasterboards that presents low moisture permeability. This suggests that the small RH differences between the techniques are mainly due to the unequal internal air temperatures rather than a specific moisture buffering effect. For a given dew point and its corresponding absolute humidity the RH will change inversely, albeit nonlinearly, with the temperature. The best summer solution, namely masonry, with lowest indoor temperatures presents the highest RH values.

2.2.7 Improving the worst solution W6

The wall W6 is the worst solution for all year round, since it has the worst dynamic thermal parameters: a very low internal areal heat capacity (and as a consequence the lowest inner inertia) and the highest decrement factor (the poorest ability to dampen the crossing heat wave). However, the accurate choice of the inner finishing layer could have great incidence on the highlighted dynamic thermal parameters.

The most appropriate intervention is to introduce an inner massive finishing that not only increases the κ_1 value (as W clay in Table 2.4, characterized by the introduction of a clay panel) but also allows a strong reduction of the crossing heat flow (W6 cement-wood). The latter intervention, that envisages the introduction of a cement-wood panel, reduces the summer inside surface temperatures down to 1°C (Fig. 2.7), with even more appreciable incidence on the operative temperatures.

Table 2.4 Dynamic parameters optimization for envelope W6 through the introduction of two types of inner massive linings

	W6	W6 clay	W6 cement-wood
Wall thickness (cm)	28	28	28
U (W/(m² K))	0.26	0.26	0.26
Y_{12} (W/(m² K))	0.043	0.041	0.022
κ_1 (kJ/(m² K))	20	32 ↑	60 ↑↑
f (−)	0.16	0.16	0.09 ↓
Max. operative temp. (°C)	31.5	30.2	30

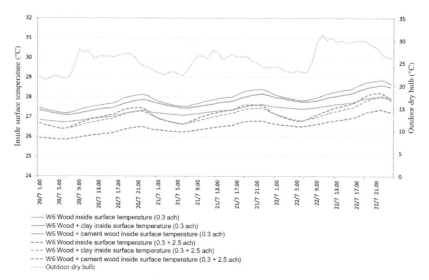

W6 Wood inside surface temperature (0.3 ach)
W6 Wood + clay inside surface temperature (0.3 ach)
W6 Wood + cement wood inside surface temperature (0.3 ach)
W6 Wood inside surface temperature (0.3 + 2.5 ach)
W6 Wood + clay inside surface temperature (0.3 + 2.5 ach)
W6 Wood + cement wood inside surface temperature (0.3 + 2.5 ach)
Outdoor dry bulb

Figure 2.7 Numerical simulations. Effect of the introduction of two different types of inner massive linings on the inside surface temperatures of envelope W6. *From Energy, comfort and environmental assessment of different building envelope techniques in a Mediterranean climate with a hot dry summer, Appl. Energy 134 (December 1, 2014) 176–196, ISSN 0306-2619.*

2.2.8 Dynamic parameters affecting the winter consumptions

The U value is the main stationary parameter affecting the heating consumptions [9]. Also the f value plays an important role in winter. The reason is well established in literature and is briefly explained.

During the winter period the great differences between external temperature (around $5-10°C$ even for the mild temperate climate of Ancona) and internal temperature (heated continuously with a $20°C$ temperature set-point) create an outgoing flux that involves the entire wall stratigraphy and all the layers of the external wall take part in the heat transfer process. As a consequence all the envelope layers contribute with their damping and thermal barrier effect. A wall characterized by a high resistance to heat transfer through the presence of multiple layers (each providing an additional thermal resistance) or the adoption a single layer built up with massive and low conductive mixed materials would be preferable for a high thermal barrier effect if the thermal transmittance U is fixed. The main characteristic that influences this effect is the decrement factor f. A wall with low f value will provide low consumptions. This dynamic thermal characteristic allows to mitigate the thermal wave coming from inside which crosses the opaque wall, thus storing the internal heat for longer period and ensuring a better performance for the wall.

The comparison between walls W4, W5, and W6 in winter (Fig. 2.8) shows that the W5 solution (wood-cement) has the best behavior for its lowest decrement factor. The increasing of the f value when the other parameters are fixed (W4—masonry) determines a downward shifting of the temperature curve, while the simultaneous reduction of κ_1 value (thus configuring the wall W6) determines higher temperature variability, as happened for walls W1, W2, and W3 (Section 2.2.4).

The winter consumptions obtained through a dynamic energy analysis for the city of Ancona with intermittent operation (Table 2.5) show a

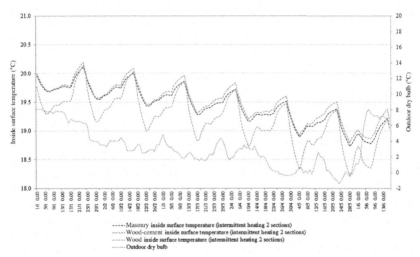

----- Masonry inside surface temperature (intermittent heating 2 sections)
----- Wood-cement inside surface temperature (intermittent heating 2 sections)
----- Wood inside surface temperature (intermittent heating 2 sections)
········· Outdoor dry bulb

Figure 2.8 Simulation results. Winter comparison between walls W4, W5, and W6 (inside surface temperatures), under intermittent heating operations.

Table 2.5 Winter consumptions for the three techniques

	Useful energy demand (kWh prim./m² year)	Average seasonal RH value (%)
W4 masonry	19.43	58.5
W5 wood-cem.	19.37	58.4
W6 wood	19.54	58.7

2 sections: 5:00 a.m.−9:00 a.m., 3:00 p.m.−11:00 p.m.
Winter: HDD = 1616; T_{med} = 9.3°C, T_{min} = −1.4°C; Δ°C = 10.3°C.
Ancona (central Italy, type *Csa*: hot-summer Mediterranean climate). No shading devices.

small difference, of about 1%, between best and worst solution, due to the same U value. The wood-cement W5 resulted the best solution in winter for its highest dampening effect on the outgoing wave (thanks to its lowest f value). For RH values the same consideration drawn for summer period could be done.

2.2.9 A proposal for superinsulated envelopes in temperate climates

Among different highly energy efficient envelopes (with very low U values), the best solution for all year round will be the one with the highest internal areal heat capacity (and as a consequence the highest inner inertia) and an appropriate damping effect on the incoming wave (a low f value, but not too much low). A detailed explanation of this fact is given below.

Low values of the decrement factor combined with high values of internal areal heat capacity and high values in the time shift of periodic thermal transmittance is known to designate the best wall configuration when the aim is to decrease the effect of the external thermal loads during summertime (Refs. [7,10]). However, as demonstrated above, very low f values configure opaque surfaces behaving as thermal barrier, especially if combined with very low U values. In this case the wall will block not only the incoming heat flux but also the outgoing one thus determining the overheating of the room. So the adoption of high internal areal heat capacity is of particular importance. This situation is still scarcely studied since the researches on the dynamic parameters are generally focused on envelopes with medium U values, given that the superinsulation is a relatively new issue and very rarely the abovementioned studies combine the deepening on the thermal dynamic properties with the identification of the relative indoor comfort conditions. Moreover very low f values are achievable only through new techniques, such as highly stratified multilayered walls or cement-wood envelopes that are a young research field.

Returning to the problem formulated in this chapter (Table 2.1), it is now clear that the wall W5 is characterized by a very low f value behaving as a thermal barrier. The strong reduction of temperature amplitude of the outgoing wave determines that the heat is trapped within the envelope internal layers and a quote of this heat is then retransmitted inside, thus contributing to internal overheating (Fig. 2.9). Walls W1, W2, W3, and W4 are all characterized by low-medium f values; this is the best situation since the dynamic behavior of the envelope is ensured and all the

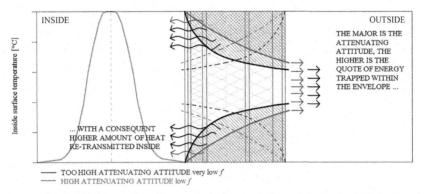

Figure 2.9 Overheating effect due to the adoption of too low *f* values. *From Super-insulated wooden envelopes in Mediterranean climate: summer overheating, thermal comfort optimization, environmental impact on an Italian case study, Energy Build. 138 (March 1, 2017) 716–732, ISSN 0378-7788.*

Table 2.6 Proposal of combined limits of dynamic thermal parameters, valid for energy efficient envelopes with $U < 0.3$ (W/(m² K))

f (−)	Internal areal heat capacity κ_1 (kJ/(m² K))	
1. $f < 0.04$ thermal barrier	$\kappa_1 \geq 50^a$	W5[b]
2. $0.04 \leq f \leq 0.08$ dynamic wall[c]	$\kappa_1 \geq 40$	W2,W3,W4
3. $0.08 < f < 0.3$ entering flux	$\kappa_1 \geq 50$	W6
4. $f > 0.3$ too high flux	$\kappa_1 \geq 60$	

[a]To be increased up to 60 for $U < 0.2$ W/(m² K).
[b]The walls are detailed in Table 2.1.
[c]Optimal solution.

layers participate to the thermal balance; in that situation medium–high κ_1 values are anyway recommended. Wall W6 is characterized by the highest *f* value, thus it has the lowest damping effect in the incoming heat flow, determining adjunctive heat gains in the internal overheated environment. So it is very important in this case to adopt very high κ_1 values, e.g., with very massive inner linings.

On the basis of these considerations, using numerous simulations it was possible to obtain combined threshold values for the internal areal heat capacity κ_1 and *f* value (Table 2.6).

The most suitable solutions in temperate variable climates are within category 2, characterized, as learned by traditional architecture, by the best interaction with the external and internal environment. Very interesting in such temperate climates could be the dynamic solutions (such as ventilated facades) that behave as thermal barriers in winter (with very low f value) but increase the same value during the summer through the "deactivation" of some external layer.

2.3 IMPACT OF THERMAL CAPACITY IN DIFFERENT DESIGN CONDITIONS

2.3.1 Problem description

The optimal envelope choice regarding the selection of materials, layers thicknesses, and the relative position of thermal resistant and massive layers, as well as the importance of making this correct choice, are strongly influenced by specific building features. The exposed building envelope area, glazed percentage, use of shadings or overhangs, presence of adjacent "buffer" spaces, occupants habits on the use of the heating plants and ventilation, all impact on envelope performance. This section is an attempt to provide quantitative data and debate on these complex interactions by comparing various situations that could occur in a multistory building.

Consider the building layout shown in Fig. 2.10. The apartments on the top floors are laid under the roof and will have major solar gains but also higher heat loss during the night. Even more the apartments with higher glazed surface will present overheating that could also vary depending on the different exposure and on the effectiveness of the natural cross-ventilation. The presence of protruding roofs that shelter the sun radiation in the upper floors or the adjacency to the ground in the lower floors acting as heat tank, could strongly influence the internal room temperature regime and as a consequence determining different impact of an optimized massive envelope.

2.3.2 Skin dominated versus core dominated

A skin dominated building has the large part of energy loads determined by the enclosure, with high solar gains and heat losses across it. Buildings with high occupants' densities and high internal heat gains (e.g., for highly glazed walls) are instead considered core-dominated buildings. Residential buildings are typically skin-dominated buildings, while

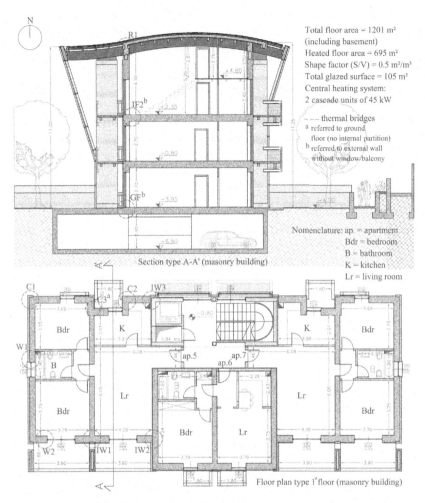

Total floor area = 1201 m²
(including basement)
Heated floor area = 695 m²
Shape factor (S/V) = 0.5 m²/m³
Total glazed surface = 105 m²
Central heating system:
2 cascade units of 45 kW

--- thermal bridges
ᵃ referred to ground
floor (no internal partition)
ᵇ referred to external wall
without window/balcony

Section type A-A' (masonry building)

Nomenclature: ap. = apartment
Bdr = bedroom
B = bathroom
K = kitchen
Lr = living room

Floor plan type 1ˢᵗ floor (masonry building)

Figure 2.10 Plan and section of simulated building. *From Energy, comfort and environmental assessment of different building envelope techniques in a Mediterranean climate with a hot dry summer, Appl. Energy 134 (December 1, 2014) 176–196, ISSN 0306-2619.*

schools are an example of the latter type of buildings. Even within the same residential building some apartments could have a behavior more similar to a skin-dominated condition, while others to a core dominated one. For example, two apartments with the same geometries, but characterized by different glazed surface (e.g., 70% of the total envelope area and 30%) will have two different behaviors. In the former there are high heat gain coming from the internal side (for the greenhouse effect), while in the latter the summer heat gains will mainly come from the opaque external wall. So while in the former case envelope techniques with high damping attitude are not recommended, in the latter case they are to be strongly encouraged.

The summer maximum operative temperatures were analyzed for each apartment of the building (Fig. 2.11), by varying the wall, W4, W5, and W6 (see Section 2.2.1). Wall W5 is that with high damping effect of the thermal wave: it will present the worst behavior in the core-dominated apartment, while it will have the best performance in skin-dominated conditions. The comparison of the three techniques at the varying of apartment characteristics demonstrated that, while the W4 solution (masonry) remains constantly the best one, the identification of the worst technique depends on the particular situation. The worst is the light-weight one (W6) for all the "skin-dominated" apartment, while it is the wood-cement one (W5) for all the "core-dominated" apartments.

The apartments on the ground floor have a core-dominated behavior since the presence of a reflecting pavement adjacent to the external enve-lope (instead of the shading balconies of the upper floors) determines the presence of high heat gains. Nevertheless, the values of seasonal peaks of temperature are not so elevated (of about 30°C in the worst condition) for the presence of a lower basement that behaves like a thermal tank. The elimination of the lower garage and the adjacent pavement that reflects the sun rays causing high internal gains in apartment 3 determines a passage from core dominated to skin dominated. As a consequence, the

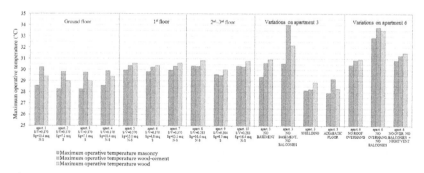

Figure 2.11 Summer maximum operative temperatures for each apartment of the building with envelopes W4 (masonry), W5 (wood-cement), and W6 (wood) and at the varying of some building features: elimination of basement, elimination of shield-ing, introduction of horizontal adiabatic floors, elimination of roof overhang, and introduction of natural ventilation.

solution W5 is no more so deleterious. Simultaneously the beneficial effect of the inferior buffer space (the basement) is reduced so that the maximum operative temperatures are higher than in the initial situation. The subsequent elimination of the shading balconies of the upper floors determines an even more overheated environment with very high internal gains and another changing of the relative preference between the three techniques. The same observations could be done for all the other simulations regarding apartment 6.

2.3.3 Shaded versus unshaded

Envelopes characterized by a significant thickness of insulating material avoid the dispersion toward the outside of the internal heat gains (coming from equipment, people, solar radiations through transparent surfaces) completely devolving to windows the thermal exchanges between indoor and outdoor. This determines the increasing importance of the design of glazing surfaces and external shading devices, which have to meet sometimes–contrasting requirements: maximizing the solar irradiation and thermal resistance during the winter and minimizing the solar gains during the summer, always guaranteeing the necessary level of natural lighting. Window shading strongly reduces the solar gains, thus changing the behavior from core dominated to skin dominated.

The summer maximum operative temperatures were analyzed for each apartment with the three walls W4, W5, and W6, introduced in Section 2.2.1, considering an environment with shaded windows (Fig. 2.12) to make a comparison with the result of the previous section obtained for

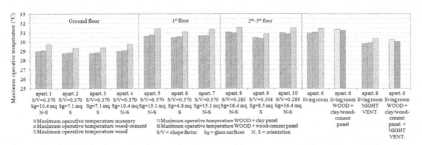

Figure 2.12 Numerical simulations. Summer maximum operative temperatures for each apartment of the building by varying the wall, W4 (masonry), W5 (wood-cement), and W6 (wood) and at the varying of some building features, in shaded condition. *From Energy, comfort and environmental assessment of different building envelope techniques in a Mediterranean climate with a hot dry summer, Appl. Energy 134 (December 1, 2014) 176−196, ISSN 0306-2619.*

unshaded glazed surfaces. The best comfort conditions are still reached for all cases by adopting a massive solution (W4), while, differently by the unshaded condition, W6 is always the worst choice regardless the specific building level for each apartment. In fact, the presence of external shadings determines that all the apartments behave as skin dominated ones.

The variations on apartment 6 regard an introduction on an internal massive lining (clay or wood-cement panel) only on the wooden envelope W6 (aspect deepened in Section 2.2.7), nocturnal ventilation, or a combination of these two strategies.

The adoption of various passive cooling measures guarantees better comfort conditions.

Regarding winter consumptions, the useful energy demands were calculated with different assumptions on radiation contributions (Table 2.7). For the initial shaded solution a real use profile of shading was simulated: open during the day (8:00 a.m.−8:00 p.m.) and closed at night (8:00 p.m.−8:00 a.m.) in winter; closed for a solar radiation greater than 200 W/m^2 in summer. Subsequently the radiation on opaque surface was excluded by changing the assumptions on absorption coefficient of solar radiation and emissivity of the external surface materials; finally a totally shaded solution was explored by simultaneously excluding gains on both transparent and opaque surfaces. As highlighted earlier, the very low

Table 2.7 Winter and summer consumptions under continuous operation, city of Ancona (central Italy)

Useful energy demand (kWh prim./m² year) continuous operation, Ancona

Constr. techniques/ analysis	No shading devices	Real use of shadings	No radiation on opaque surfaces	Totally shaded
Winter				
W4: masonry	22.68	22.64	24.03	30.24
W5: wood-cem.	22.62	22.58	23.99	30.15
W6: wood	22.86	22.83	24.21	30.21
Summer				
W4: masonry	7.72	7.43	6.21	2.35
W5: wood-cem.	7.80	7.50	6.25	2.35
W6: wood	7.92	7.62	6.44	2.63

Source: F. Stazi, E. Tomassoni, C. Bonfigli, C. Di Perna, Energy, comfort and environmental assessment of different building envelope techniques in a Mediterranean climate with a hot dry summer, Appl. Energy 134 (December 1, 2014) 176−196, ISSN 0306-2619.

difference between the techniques energy demand, is due to the same *U* value. In the totally shaded situation the internal environment does not have benefice from the solar gains in winter and has no overheating phenomena in summer; the low differences between the techniques are strongly reduced down to zero and the passage from a core dominated to a skin dominated environment make the wood–cement solution the most suitable one throughout the year.

2.3.4 Continuously used versus intermittently used

The analysis of the inside surface temperatures of a southern wall (living room of apartment 6), during the coldest winter week (December 15–21), with continuous heating (Fig. 2.13, *continuous lines*), shows that the massive solutions (masonry and wood–cement) present higher and more stable values than the wooden one. The wood, for its worst dynamic thermal parameters, responds more quickly to the external variations producing slightly greater fluctuations of the daily inside surface temperature and causing less winter comfort. The intermittent operation (Fig. 2.13, *dashed lines*), although beneficial for energy saving, worsens the

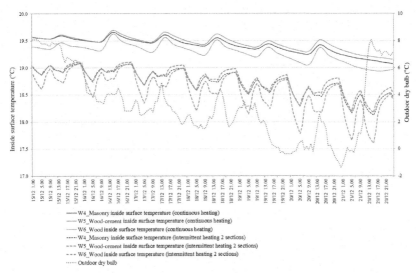

Figure 2.13 Numerical simulations. Incidence of heating profiles on the behavior of the three envelopes. *From Energy, comfort and environmental assessment of different building envelope techniques in a Mediterranean climate with a hot dry summer, Appl. Energy 134 (December 1, 2014) 176–196, ISSN 0306-2619.*

interior comfort level lowering the temperature up to 1°C for all techniques. Nevertheless, in the coldest days with intermittent pattern the efficacy of higher inertia solutions becomes more relevant because,
differently from the lightweight solution, they maintain high temperatures
also during switch-off times.

By changing the intermittent profile (2, 3, 4 sections) for the three
techniques, it is possible to note (Fig. 2.14) that a daily activation with 4
time slots guarantees the better indoor conditions with more elevated
temperature for all techniques (especially for wood-cement), even if it

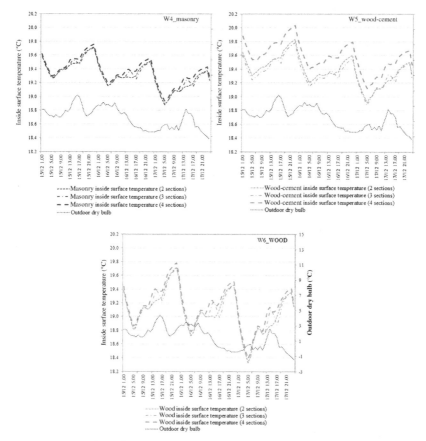

Figure 2.14 Detail on the incidence of different heating timetables (2, 3, 4 sections)
for the three analyzed techniques. *From Energy, comfort and environmental assessment of different building envelope techniques in a Mediterranean climate with a hot
dry summer, Appl. Energy 134 (December 1, 2014) 176–196, ISSN 0306-2619.*

requires slightly higher consumptions respect to the other switching profiles.

Regarding consumptions (Table 2.8) comparing a continuous use pattern of the heating plant with an intermittent use (e.g., 2 daily sections) it is possible to note that in every construction technique the winter consumptions decrease to about 14% and the summer ones to about 24%. Moreover the differences between the three techniques are slightly reduced also confirming the worst and best solution obtained with continuous use. Varying the intermittence time slots to a more fractionated daily use of the system (3—4 sections) both winter and summer consumptions increase. This happens because of the more ignitions of the system with consequent energy expenditure. During the summer it has also been hypothesized a nighttime cooling (with 4 sections profile) since (unlike the winter period) there are not regulations that limit the summer daily

Table 2.8 Winter and summer consumptions under continuous and intermittent operation, city of Ancona (central Italy)
Useful energy demand, intermittent operation,[a] Ancona (central Italy, type *Csa*: hot-summer Mediterranean climate)

Construction techniques	Continuous	2 sections[a] (kWh prim./ m² year)	3 sections[b] (kWh prim./ m² year)	4 sections[c] (kWh prim./ m² year)	
Winter					
W4 masonry	22.68	19.43	19.51	19.62	
W5 wood-cement	22.62	19.37	19.51	19.54	
W6 wood	22.86	19.54	19.67	19.73	
Summer					
W4 masonry	7.72	5.86	5.96	6.12	$(4s_{night}{}^{d} = 6.05)$
W5 wood-cement	7.80	5.93	6.03	6.17	$(4s_{night} = 6.11)$
W6 wood	7.92	6.03	6.15	6.28	$(4s_{night} = 6.24)$

[a]*2 sections*: 5:00 a.m.—9:00 a.m., 3:00 p.m.—11:00 p.m.
[b]*3 sections*: 5:00 a.m.—9:00 a.m., 12:00 p.m.—2:00 p.m., 5:00 p.m.—11:00 p.m.
[c]*4 sections*: 5:00 a.m.—9:00 a.m., 11:00 a.m.—1:00 p.m., 3:00—5:00 p.m., 7:00 p.m.—11:00 p.m.
[d]*4 sections night*: 7:00 a.m.—11:00 a.m., 2:00 p.m.—4:00 p.m., 7:00 p.m.—11:00 p.m., 2:00 a.m.—4:00 a.m.
Source: F. Stazi, E. Tomassoni, C. Bonfigli, C. Di Perna, Energy, comfort and environmental assessment of different building envelope techniques in a Mediterranean climate with a hot dry summer, Appl. Energy 134 (December 1, 2014) 176—196, ISSN 0306-2619.

system activation. The distribution of one of the 4 sections during the night, even just for 2 hours, cools the mass thus ensuring less consumptions even if the 2 sections profile is still to prefer.

2.3.5 Ventilated versus unventilated

The introduction of night ventilation strongly reduces the differences between the techniques. The comfort level of the living room (apartment 6 in Fig. 2.10) in a hot summer (city of Ancona) was evaluated, according to the adaptive comfort model (Fig. 2.15) with different natural ventilation profiles. The following conditions were set: a continuous profile set to 0.3 air changes per hour (ach); a variable profile that provides 0.3 ach during the day (6 a.m.−6 p.m.) and night ventilation with 2.5 ach (6 p.m.−6 a.m.).

The graph shows that with medium regime of air changing throughout the day, the masonry W4 and wood-cement W5 techniques, guarantee the best comfort conditions showing a very similar trend. The operative temperature of the wood apartment (W6) presents major fluctuations and higher discomfort peaks during the hottest period. The use of high rates of natural ventilation during the night decreases the temperatures bringing all the curves closer to optimal temperature but in the same

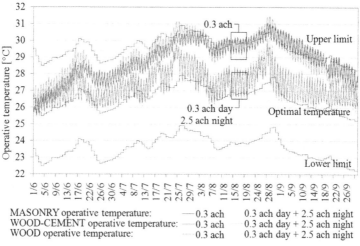

Figure 2.15 Operative temperatures: comparison between envelopes W4, W5, and W6 with and without ventilation. *From Energy, comfort and environmental assessment of different building envelope techniques in a Mediterranean climate with a hot dry summer, Appl. Energy 134 (December 1, 2014) 176−196, ISSN 0306-2619.*

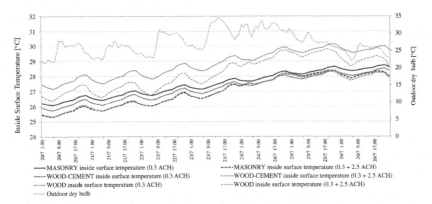

Figure 2.16 Surface temperatures: comparison between different envelopes W4, W5, and W6 with and without ventilation. *From Energy, comfort and environmental assessment of different building envelope techniques in a Mediterranean climate with a hot dry summer, Appl. Energy 134 (December 1, 2014) 176–196, ISSN 0306-2619.*

time, it increases the daily temperature fluctuations reducing the difference among the three techniques.

The analysis of the surface temperatures (Fig. 2.16) in the summer hottest week (July 20–26) confirms the better performance of both massive solutions (masonry W4 and cement-wood W5) that ensure lower (about 1°C) and more stable surface temperatures than the wooden wall (W6). The night ventilation (*dashed lines*) reduces the surface inside temperatures for all techniques of almost 1°C and reduces the difference among the technologies.

 ## 2.4 THE IMPORTANCE OF THE OCCUPANTS' BEHAVIOR

2.4.1 Problem description

Typical example of a core-dominated building could be the social housing buildings, characterized by high density of occupation, limited use of heating plants, scarce ventilation; these combined factors could promote high indoor humidity levels and the risk of condensation. As a consequence retrofit interventions on such buildings, through the introduction of additional insulation layers, as required by the regulations, may cause summer overheating phenomena [11] or be disadvantageous from global costs evaluations for the limited use of the plants.

Studies on the overall cost of different retrofit interventions were performed in literature [12,13] with the aim of developing methods and procedure of comparative assessment of intervention combinations. In general the real behavior of the walls (through in-situ measurements) and the real users operations are not considered. Some authors [14] addressed this issue through on-field monitoring with the aim of demonstrating that the changing of the attitudes of the occupants (called "human-based retrofit") could be an effective alternative to traditional insulation interventions.

Only recently some research study was focused on the prediction of the real behavior of the occupants in order to verify the real effectiveness of alternative envelope solutions from the point of view of energy saving, comfort, and global costs.

The present section reports the results of a monitoring campaign on a social housing building (Fig. 2.17, building B5 in Appendix A), aimed at evaluating the influence of occupants' behavior on internal temperatures and humidity and on the efficacy of retrofit interventions.

2.4.2 Occupants' behavior on the use of the heating plants

The indoor conditions are closely connected to the specific use of the system.

Fig. 2.18 shows the operative and surface temperature recorded during the experimental survey in the living room of different apartments. The temperatures, with the exception of the first floor apartment above the porch (U1), present very similar trends with values ranging from a minimum of about 17°C to a maximum of about 20°C. The similarity is due to the fact that these units are all characterized by an intermittent use of the heating system and a limited number of operating hours (U2 = 8 h, U3 = 7 h, U4 = 4 h, U5 = 6 h). In the case of the first floor (apartment U1) a frequent ignition for the system (12 h) and the setting of a high value for temperature set back (18.5°C) ensure high operative temperatures, with minimum swings and values constantly between 18.5°C and 20.5°C.

The heat flows (*gray area* of the graph) are as expected always outgoing with values between 5 and 15 W/m^2.

U1   First floor above an unheated space, double exposure.

U3   Third floor, double exposure.

U3$_{corner}$ Third floor, double exposure on the building corner.

U3$_{single}$ Third floor, single exposure.

U5   Fifth floor bordering on the roof, double exposure.

U$_{blower}$ Fifth floor bordering on the roof, single exposure.

Figure 2.17 Social housing building in Ancona: view from the eastern side, drawings of the eastern elevation and plan with the identification of different apartment units. U$_{blower}$ is an apartment in which blower door test was carried out after the introduction of new sealing stripes in window frames.

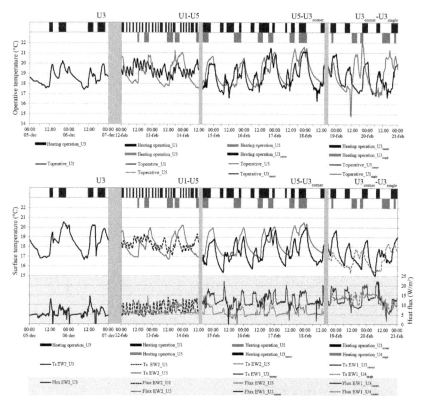

Figure 2.18 Experimental results. Winter operative temperature recorded in the living room in different apartments. The bars in the upper part of the graph indicate the hours in which the heating plant was switched on.

The comfort evaluation through the predicted mean vote (PMV) method (Fig. 2.19) shows that all apartments have poor comfort conditions for low temperatures.

To generalize the results to the entire winter season extensive simulations were performed for all the heated period (from November 1 to April 15). Starting from the appropriately calibrated models (through comparison with measures) a new model including one accommodation for all the building levels was simulated by imposing the internal gains according to relative standards (see Appendix B, subsection B2, Step 2) and varying three heating programs (24, 7, 4 hours).

Table 2.9 shows the winter consumptions calculated for each apartment under dynamic regime in the reference earth-sky model (with standard conditions) with the three different ignition programs. The table also reports the consumptions relative to the models calibrated with the real

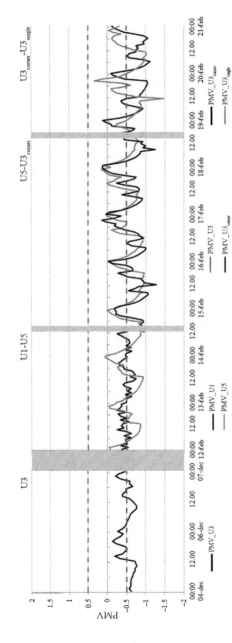

Figure 2.19 PMV evaluation for different apartments. *Data from experimental measures.*

Table 2.9 Winter consumptions (November 1–April 15) of the different apartments

Winter consumptions (kWh/m² year)	U1	U2	U3corner	U3single	U3	U4	U5	Ublower
Apartment units simulated in *real usage* conditions	49.7	—	29.3	12.5	18.8[a]	—	52.5	20.8[b]
(Effective hours of heating on)	(12)[c]	—	(7)	(4)	(6)	—	(8)	(3)
Apartment units simulated in *standard usage* conditions								
Hours of heating operation:								
24 h (semi-steady conditions)	49.4	20.5	—	—	20.5	20.5	66.5	—
24 h (dynamic conditions)	51.3	20.7	—	—	20.5	21.4	58.0	—
7 h	43.1	19.9	—	—	18.8	20.7	47.4	—
4 h	32.0	17.2	—	—	16.4	17.9	35.1	—

[a]21.3 kWh/m² year real consumptions derived from the bills (November 2009–April 2011).
[b]19.4 kWh/m² year real consumptions derived from the bills (November 2009–April 2011).
[c]The values in brackets are the sum of hours in which the occupants of the relative apartments have effectively switched on the heating plants (average values obtained during the survey).
Source: Data from experimental survey and from analytical simulations.

usage conditions (to make a comparison with the standard assumptions), the consumption obtained in semistationary regime, and those derived from the survey of the real consumptions for two apartments (U3 and U_{blower}). The comparison of the results shows good reliability of the simulation models. This is particularly evident by comparing the consumptions obtained for models calibrated with the real conditions of use and the values derived from the bills (on the footnote of the table). The ground floor apartment U1 and that under the roof U5 are the most disadvantaged ones regardless of the heating program considered.

The reduced heating profile determines high energy saving only for the more disadvantaged apartments, on top floor and above the porch for the high heat loosing surfaces. However, this would lead to unacceptable thermal discomfort levels.

2.4.3 Different use of natural ventilation and envelope air permeability

In the heating season when the aim is to minimize ventilation heat loss, the occupants' behavior regarding windows opening has a noticeable effect on comfort levels. However, daily air exchange is very important for occupant's well-being and to control RH regimes.

Fig. 2.20 shows the winter behavior of flats at the varying of envelope permeability, mode of use of the heating plants, and occupants' habits in cross-ventilation activation.

Four days with similar external climatic conditions were chosen (from December 12 to 13 and from February 6 to 7) to allow horizontal comparisons.

The figure on the left shows the comparison between two units, U3 and U_{blower}. They are characterized by a similar use of the heating (very rarely switched on) but have different levels of air tightness, greater in the case of U_{blower} that provided new sealing for external window frames, and a different use of the natural ventilation, present in the first apartment U3 and almost absent in the second. The *solid lines* (*black* and *gray*, respectively) show similar air temperatures, with a maximum of 20°C and a minimum of $18-16$°C. This result is consistent with the monitored consumption values that for the two apartments are very similar (approximately 20 kwh/m^2 year). Even with similar temperature conditions, the different envelope permeability and the unequal occupant' habits in the activation of natural ventilation determine very different internal RH levels (around 50% for U3 and 70% for U_{blower}).

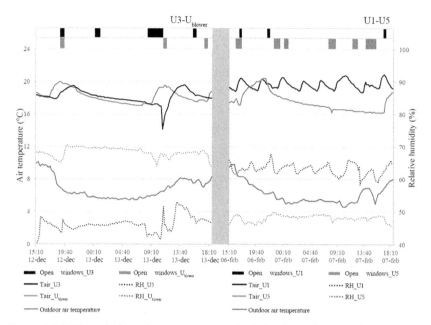

Figure 2.20 Winter indoor air temperature and relative humidity recorded in the living room in different apartments (data from experimental measures). The bars in the upper part of the graph indicate the hours in which the windows were open.

The figure on the right shows the comparison between two units (U1 and U5) characterized by a different use of the heating (continuous in U1 and intermittent in U5) and a different use of natural ventilation (almost absent in the first case and present in the second). U1 (*black line*) is used in a conservative manner, while U5 (*gray line*) in a more dynamic way, with more frequent windows openings. The graph shows that the unit on the first floor is characterized by the highest temperatures for continuous use of the system, while the unit at the top floor shows a very similar trend to that recorded in the previously analyzed units. The RH levels (65% on the first floor and 50% on the fifth) are closely influenced by the different air exchange in the two cases.

2.4.4 Global convenience of an intervention of superinsulation

The study of the global convenience of different retrofit interventions (see details on Section 3.6.3) with different time slots for the heating plants (24, 7, 4 hours) demonstrates that the extreme reduction of the ignition profile (as in the real conditions for the present social housing building chosen as example) nullifies the convenience of any type of intervention (Fig. 2.21).

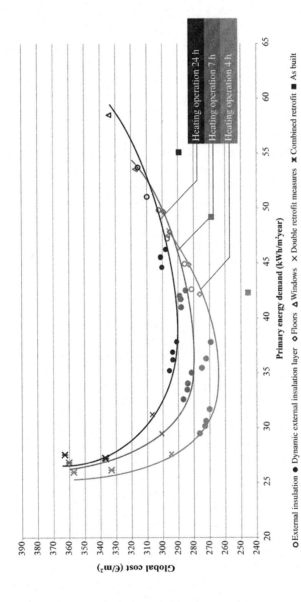

Figure 2.21 Global cost evaluation for different retrofit measures at the varying of the heating operation.

The curves represent the relationship between global costs and energy demand of alternative retrofit intervention of the building. The minimum of each curve represents the optimum intervention.

The decrease of time of use of the plants determines that the curve of the global costs tends to be shifted downward for the reduction of the overall cost (due to the decrease in running costs for heating) and it moves toward the origin of the axes for the corresponding decrease of the primary energy demand. However, the most important changes are the distance between the minimum of the curve and the "as built" initial situation (represented as a squared point). At the reduction of heated hours the vertical detachment strongly increases.

With a continuous heating the global costs of the initial solution is similar to those of the retrofitted solution, even if the energy demand is strongly reduced: so the investment result to be convenient. With a heating operation of 7 hours and down to 4 hours the overall cost of the retrofit is very higher than that of the "as built" scenario (difference of about 10 €/m^2 for operation of 7 hours and about 20 €/m^2 for 4 hours) for the higher incidence of initial investment than the costs reduction achieved in the use phase.

2.5 THE COMPLEX INTERACTION BETWEEN MASS AND OTHER FACTORS

2.5.1 Problem description

The present section reports an experimental evidence on the complex interaction between the mass and other factors such as seasonal climate variations, ventilation, and heat losses through other components. Moreover the "trend inversion" between walls with different mass on a vernacular building in a temperate climate is clearly visible through measures.

The selected building (Appendix A, case study B.3) has a massive envelope characterized by different inertia at the two building levels (Fig. 2.22), namely a three-wythe solid brick masonry at the ground floor and semisolid bricks at the first floor. They present a similar high level of internal areal heat capacity, near 60 kJ/(m^2 K) for both envelopes, but two different decrementing attitude of the thermal wave, higher for the heavy masonry at the ground floor.

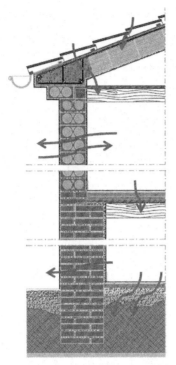

Figure 2.22 Section of a building with two types of envelopes at the two levels and multiple interactions with the other building components.

2.5.2 Trend inversion

In summer (Fig. 2.23) the external surface temperatures reach their minimum value at around 5:00 a.m. in both walls with lower values (about 2°C) recorded in the first floor lightweight wall. In the following hours, with the solar radiation rising, the external surface temperatures increase, with the same trend for the two walls (since they have the same external plaster finishing) reaching the maximum at about 11:00 a.m., with higher values (about 2°C) for the low inertia wall. The difference could be ascribed to the fact that, in this wall, a greater fraction of heat absorbed on the external side is transmitted in the adjacent layers (for its higher decrement factor).

The internal surface temperatures show different fluctuations at the two levels with a maximum daily range of about 4°C for lightweight wall and 1.5°C for the massive ones. Moreover, the two curves have a different slant. The massive wall surface temperature increases slowly and the

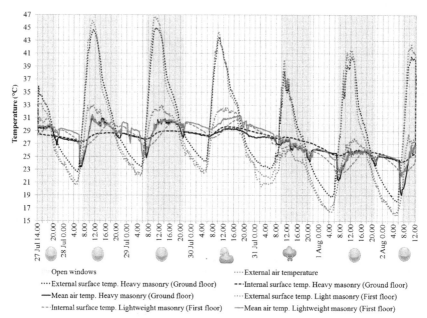

Figure 2.23 Summer experimental measures on the ground floor and on the first floor of building B3 (Appendix A). *From The effect of high thermal insulation on high thermal mass: is the dynamic behaviour of traditional envelopes in Mediterranean climates still possible?, Energy Build. 88 (February 1, 2015) 367–383.*

maximum value is kept for a long time (about 12 hours: from 11.00 a.m. to 00.00 p.m.); the low inertia wall surface temperature rises more quickly and, as soon as the maximum value was reached (about 7 hours after recording the maximum value on the outside surface), it suddenly decreases.

In autumn (Fig. 2.24) the external surface temperature of the heavy wall presents maximum values of about 4°C higher than the low inertia wall (except in rainy and cloudy days in which the temperatures are nearly equal) showing a different behavior than that recorded during the summer, when the maximum value was higher for the lightweight wall. At night, however, the behavior of the two walls is unchanged compared to the summer monitoring, with minimum values lower for the light-weight wall (about 2°C). Solid brick masonry shows internal surface temperatures higher than the semisolid brick wall, with an opposite behavior respect to the summer. There is still a greater stability of massive wall temperatures (daily temperature range of about 2°C) with respect to the lightweight wall (diurnal temperature variation of about 4°C). The

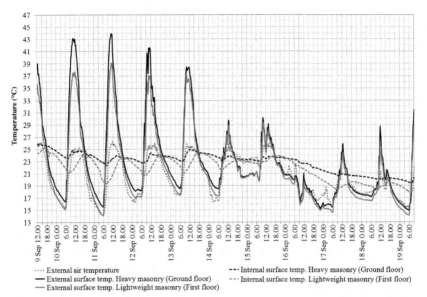

Figure 2.24 Autumn experimental measures on the ground floor and on the first floor of building B3 (Appendix A). *From The effect of high thermal insulation on high thermal mass: is the dynamic behaviour of traditional envelopes in Mediterranean climates still possible?, Energy Build. 88 (February 1, 2015) 367–383.*

different behavior of the two walls is due to the specific capability of preserving the summer stored heat and the different response to the seasonal variations.

So the solution with higher global inertia (lower f value) has a favorable behavior in both periods, with lower temperatures in the central hours of the summer days and higher values as the winter begins. A period of trend inversion for the techniques could be identified at the end of the summer, thus confirming what stressed in Section 2.2.6.

2.5.3 Impact of the mass with and without natural ventilation

The impact of natural cross-ventilation on walls with different thermal inertia is clearly visible by comparing July 29 (open windows) and July 30 (closed windows) in Fig. 2.23.

When the windows are open (*red hatched area* (gray in print versions)) the internal air temperatures (continuous lines) at two levels are equal because of the inlet of outside air. The values instead differ with closed windows. During the night, the air temperature values are higher than in the open configuration, reaching 28°C at the ground floor and 29°C at the first floor. During the day, at the ground floor low temperatures are maintained (29°C) so lowering the values of about 2°C respect to a vented ground floor, differently in the upper floor there is a thermal over-heating (about 1°C) respect to open configuration. This difference depends on the radiative contribution of the other constructive elements that are much reduced with open windows while causes overheating at the first floor and overcooling at ground floor with closed windows.

2.5.4 Impact of the presence of other heat losing elements

A dynamic simulation, starting from calibrated model through the measured data on the two floors, was carried out by placing the two walls at the same building level (ground floor) to assess how much the dissimilarities recorded at the two building levels are related to the different boundary conditions. In a subsequent variation, the heat flow through the ground floor was also eliminated by imposing an adiabatic layer in order to make the result independent from the selected story and to highlight the contribution due solely to the different envelope masses. Internal loads programs have been used according to the standard recommendations (Italian standard UNI-TS 11300-1) and a typical summer day (July 29) was chosen for the evaluation also varying the windows opening (always open or always closed).

The study of air temperatures at the ground floor (Fig. 2.25A) confirms what founded with measures in the as built situation, in which the closing of windows determines a reduction in the fluctuations of the air temperatures with lower daily values and higher nighttime values than in the naturally vented environment. Nevertheless, the low nocturnal values in the vented room, combined with the storing effect of the two walls (slightly higher for the massive one), determine that the surface temperatures are lower for the open configuration than for the closed one through the day (Fig. 2.25B and C). The internal surface temperatures are only slightly influenced by windows opening or closing for both the massive wall and the lightweight one because of the great incidence of the ground floor heat dispersions.

For both walls the introduction of an adiabatic ground floor causes the curve upward translation of 2.5°C when the windows are open (*black dotted line*) and an overheating until to 3.5°C with closed windows (*dashed*

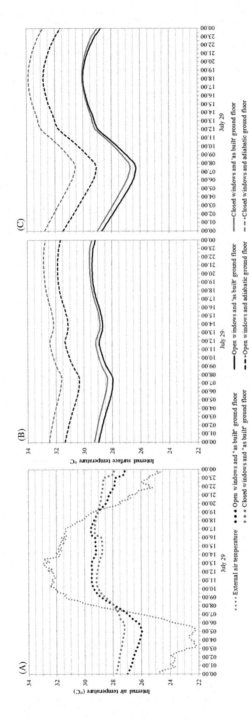

Figure 2.25 Simulations on ground floor levels of building B3 (see Appendix A) by alternatively considering the windows closed or open and by adding an adiabatic layer in the ground floor slab. (A) Ground floor air temperatures, (B) surface temperatures wall with high thermal inertia, and (C) surface temperatures wall with low thermal inertia. *From The effect of high thermal insulation on high thermal mass: is the dynamic behaviour of traditional envelopes in Mediterranean climates still possible?, Energy Build. 88 (February 1, 2015) 367–383.*

gray line). The closing of the windows determines slightly higher surface temperatures on lightweight envelope for its lower inertia.

A different fluctuation due mainly to the different inertia is highlighted by comparing the two walls temperature trends.

2.5.5 Effect of superinsulation in envelopes with different mass

The introduction of an external superinsulation (see detailed parameters in Section 3.4.1) has a different effect on the two envelopes (Fig. 2.26). While analyzing the thermal parameters of retrofitted envelopes (Table 3.2), the reader should note that the insulation has different effect at the two building levels: the internal heat capacity κ_1 is lowered in both cases (mostly for semisolid bricks) and both the two superinsulated walls have a strong reduction of decrement factor f. Only at the ground floor it drops below the value of 0.04, identified in Section 2.2.5 to be the limit below which the walls behave as thermal barriers. This is the main cause of the overheating (up to 3°C higher respect to the uninsulated wall) caused by the thick insulation especially at this building level.

The high internal mass (high κ_1) ensures a behavior more stable than adopting a new lightweight building (e.g., wall W6 in Table 2.1).

2.6 THERMAL MASS AND EXTREME CLIMATES

2.6.1 Problem description

A lot of traditional buildings born in very different climates elaborated complex thermal solutions to enhance the environmental comfort. Rainfall, humidity, and thermal conditions vary greatly with geographical location (Fig. 2.27) and have been the main determinant of traditional form of buildings.

Indeed traditional buildings born in very far places and different cultures show a substantial similarity in the constructive response to the same type of climate.

In hot dry climates with clear skies the cool air at night is able to dissipate the heat stored by the highly massive envelopes during the day, through convection and radiation. The external insulation on this kind of envelope could eliminate this beneficial cooling effect thus determining

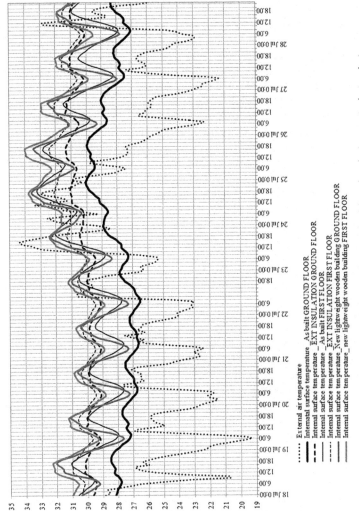

Figure 2.26 Effect of introducing an external insulation layer at the two building levels (numerical simulations on a model calibrated through measures).

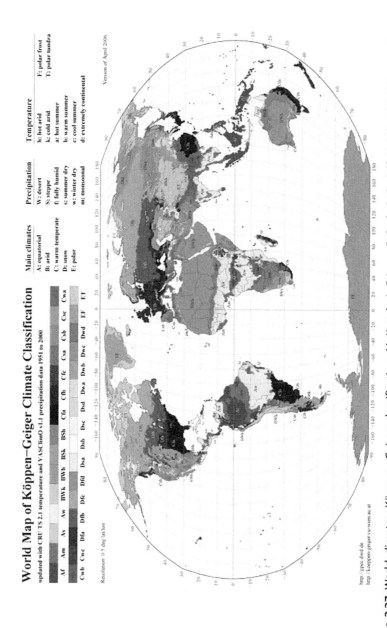

Figure 2.27 World climates, Köppen–Geiger classification. M. Kottek, J. Grieser, C. Beck, B. Rudolf, F. Rubel, *World map of Köppen-Geiger climate classification, Meteorol. Z., 15 (2006) 259–263; http://koeppen-geiger.vu-wien.ac.at.*

internal overheating. In temperate climates characterized by high weekly and seasonal temperature range this cooling benefit is even more evident. For example (Table 2.10 and Fig. 2.28) the city of Ancona (central Italy) presents winter temperatures similar to London and summer ones in some day very near to Cairo. The main difference with the latter is a major weekly fluctuation with peaks down to 15°C (even in the hot season), very positive for the mass cooling. In cold climates, as in the alpine climate of Bolzano (northern Italy), the main problem is no more the summer overheating but the heat conservation. Here lightweight fabrics (wooden envelopes) with highly insulating capability, combined with internal massive fireplaces could provide comfortable conditions. As a consequence in hot dry climates the summer comfort is a priority and high mass should be preferred while its contribution could be less appreciable in extremes climates because it does not completely play its dynamic role for the too hot or too cold daily temperatures. Moreover the superinsulation of the mass could have different effect according to the climates.

The following section deepens these aspects and highlights the new concept of "technique trend inversion."

In all the climates the following situation occurs:

1. During the winter lightweight and massive techniques (characterized by very low and equal U values) behave in a very similar way;
2. Just after the end of winter period and during the whole summer the temperature for lightweight solutions are higher than massive ones;
3. The seasonal inertia of the massive solutions determines that at the end of the hot season their temperature curve maintains its constant trend with elevated temperatures, while the lightweight technique temperature suddenly drops thus recording values always lower than massive solutions. The behavior of such lightweight solutions is disadvantageous since they show overheating in summer central period and approach the beginning of the cold season with lower temperatures. The point of curves intersection identifies a trend inversion. To make effective comparisons between different techniques the calculation of the discomfort hours for overheating in the summer should be subdivided in two periods, before and after the trend inversion.

The comparison between massive and lightweight techniques for temperate climates (with the identification of the trend inversion) was already

Table 2.10 Main thermal parameters of different studied climates

	Temperate climate		Winter extreme climates		Summer extreme climates	
Climate	**Ancona** Type *Csa*: hot-dry summer Mediterranean climate		**Bolzano** Type *Dfb*: cold temperate climate with warm summer	**London** Type *Cfb*: humid temperate climate with warm summer	**Palermo** Type *Csa*: hot-summer Mediterranean climate tending to semiarid	**Cairo** Type *BWh*: hot desert climate
	Summer	**Winter**				
Main climatic parameters[a]	CDD = 742 T_{med} = 23.3°C T_{max} = 36.9°C Δ°C = 11.7°C	HDD = 1616 T_{med} = 9.3°C T_{min} = −1.4°C Δ°C = 10.3°C	HDD = 2913 T_{med} = 3.9°C T_{min} = −11.6°C Δ°C = 13.9°C	HDD = 2866 T_{med} = 8.1°C T_{min} = −5.9°C Δ°C = 9.9°C	CDD = 1022 T_{med} = 23.8°C T_{max} = 34°C Δ°C = 7.7°C	CDD = 1746 T_{med} = 25.2°C T_{max} = 43°C Δ°C = 12.9°C
Length of season	June 1– September 30 (121 days)	November 1– April 15 (196 days)	October 15– April 15 (182 days)	September 15– June 15 (273 days)	May 15– October 31 (169 days)	April 15– November 30 (229 days)
RH ext	58.8%	68.2%	70.7%	81.2%	73.3%	56.7%

[a]Heating/cooling degree days calculated from weather file (temp. base 18°C); T_{med} and $T_{min/max}$: medium and minimum/maximum outdoor dry bulb temperature; Δ°C: mean diurnal temperature range.

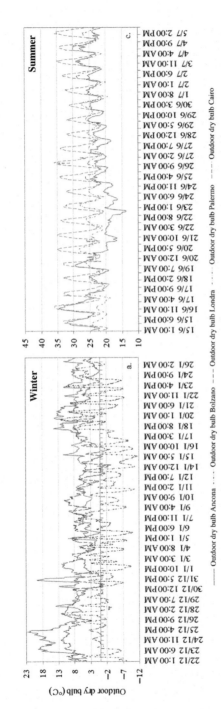

Figure 2.28 Outdoor dry bulb temperatures for a temperate climate of Ancona (central Italy) in winter and summer compared with extremely cold climates in winter (Bolzano and London) and with extremely hot climates in summer (Palermo and Cairo). *From Energy, comfort and environmental assessment of different building envelope techniques in a Mediterranean climate with a hot dry summer, Appl. Energy 134 (December 1, 2014) 176–196, ISSN 0306-2619.*

shown in Section 2.2.6. In the present section the day in which the operative temperatures of the lightweight solutions become colder at the summer ending is identified for other extreme climates.

2.6.2 Extremely hot climates

The 1st September is the day in which a trend inversion occurs between the three techniques W4, W5, and W6 for Ancona (see Section 2.2.6). For other climate zones, as shown in the following graphs, the trend inversion between the same techniques, is around 3rd September for Palermo (Fig. 2.29) and 26th September for Cairo (Fig. 2.30).

The discomfort hours in Table 2.11 are subdivided (values in parenthesis) into discomfort hours calculated in the hot season and discomfort hours falling after the trend inversion (end of the summer). The results show that generally the massive techniques have less discomfort hours for all locations in the hottest period of the summer. The wooden solution has always the worst behavior in the first part of the summer (until the day of trend inversion between the three techniques), while it records the lower overheating hours in the following period.

From the table it is possible to deduce that in the central summer period the mass has a greater incidence in the city of Palermo rather than Cairo, reducing the discomfort hours from 986 to 703 for Palermo and from 2852 to 2620 for Cairo, respectively corresponding to a decrease of 11% and 6% on total hot season hours. The major daily peaks for the latter extremely hot climate (see Fig. 2.28) determine that the overheated internal mass is not able to completely release the stored heat toward the external side during the night. So the envelopes with high and low mass have a more similar behavior.

The city of Ancona, differently by the other two selected extremely hot climates, is characterized by seasonal temperature variability (Fig. 2.28). The presence of extensive periods with low temperatures (from 20°C down to 15°C) doesn't succeed to exploit completely the dynamic behavior of the mass. In such period (from June 20 to 30) there are very small differences in the operative temperatures of buildings with the three techniques (Fig. 2.31). As a consequence the adoption of a masonry solution rather than wood one allows a reduction of about 188 discomfort hours (473−285), corresponding to about 9% on total summer season.

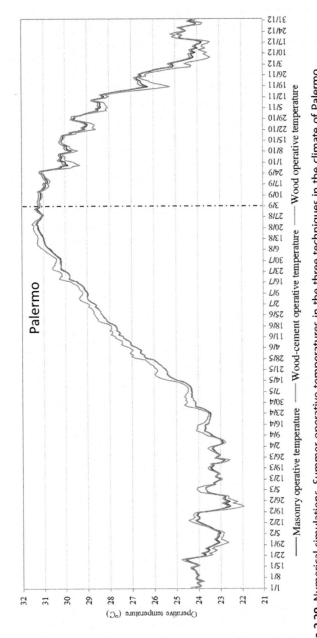

Figure 2.29 Numerical simulations. Summer operative temperatures in the three techniques in the climate of Palermo.

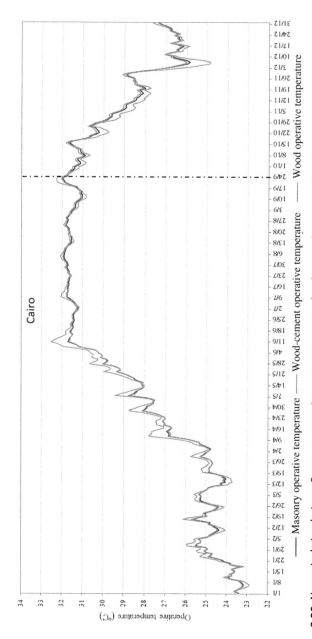

Figure 2.30 Numerical simulations. Summer operative temperatures in the three techniques in the climate of Cairo.

Table 2.11 Total overheating hours, cooling consumptions, and RH indoor values in the summer season for the three techniques W4, W5, and W6

Envelope techniques	Ancona June 1– September 30	Palermo May 15– October 31	Cairo April 15– November 30
Hours of overheating % Hours on total			
W4_masonry	643 (285 + 358)[a] 22.0%	2074 (703 + 1371) 50.8%	3995 (2620 + 1375) 72.4%
W5_wood-cem.	745 (328 + 417) 25.4%	2107 (727 + 1380) 51.6 %	4041 (2631 + 1410) 73.2%
W6_wood	695 (473 + 222) 23.7%	2231 (986 + 1245) 54.7%	4024 (2852 + 1172) 72.9%
Cooling consumptions intermittent operation, 2 sections (5:00 a.m.–11:00 a.m., 3:00 p.m.–11:00 p.m.); kWh prim./m²year			
W4_masonry	5.86	14.17	22.98
W5_wood-cem.	5.93	14.26	23.11
W6_wood	6.03	14.38	23.20
Average seasonal RH value (%)			
W4_masonry	72.9	81.0	72.2
W5_wood-cem.	72.8	80.9	72.1
W6_wood	72.7	80.7	72.1

[a]Subdivided in hours before and after trend inversion.
Source: F. Stazi, E. Tomassoni, C. Bonfigli, C. Di Perna, Energy, comfort and environmental assessment of different building envelope techniques in a Mediterranean climate with a hot dry summer, Appl. Energy 134 (December 1, 2014) 176–196, ISSN 0306-2619.

Table 2.11 also reports the mean seasonal indoor RH values of the three techniques for summer period for different localities. The results show similar indoor humidity levels for the three techniques also at the varying of the climate, regardless the very different external humidity. This suggests that the RH is more affected by indoor conditions and the small differences between the techniques are mainly due to the unequal indoor temperatures. RH is the water vapor content in a given volume of air relative to its content at saturation at the same temperature. Warmer air will hold more vapor at saturation than colder air, for higher vapor solubility under more elevated temperatures. Thus with a fixed amount of water vapor, the warmer air will record lower RH values.

The best summer solution, namely masonry, with the lowest indoor temperatures presents the highest RH values in all climates.

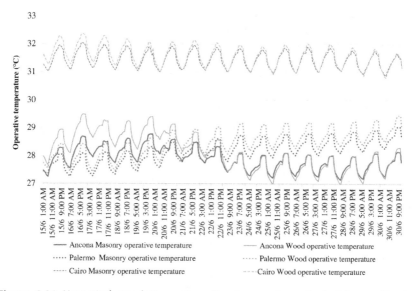

Figure 2.31 Numerical simulations. Operative temperatures in buildings with the three techniques in different climates in a typical summer week. *From Energy, comfort and environmental assessment of different building envelope techniques in a Mediterranean climate with a hot dry summer, Appl. Energy 134 (December 1, 2014) 176–196, ISSN 0306-2619.*

2.6.3 Cold climates

During the winter moving from the temperate climate of Ancona to other colder climates such as London and Bolzano, the adoption of the mass becomes less important. For example (see Table 2.12) the adoption of a wood-cement solution rather than a wood one allows a reduction of 53 discomfort hours of overcooling for Ancona (195–142) and 16 for Bolzano (809–793).

To show in detail this different behavior a graph with the internal operative temperatures for both wood-cement and wooden envelope in the three climates (Fig. 2.32) is also reported. The graph demonstrates that the different contribution given by the mass is due to a different seasonal climate variability of the considered locations. Bolzano (Fig. 2.28) is an extreme alpine climate with a constant daily temperature variation of about 10°C (mostly between −5°C and 5°C). Similarly London is constantly between 0°C and 10°C. Ancona has a very much fluctuating behavior, with temperatures between 5°C and 20°C and peaks up to 22°C. In this milder periods (e.g., between December 22 and 26 and from January 19 to 22) the beneficial effects of the mass could be better appreciated.

Table 2.12 Total overcooling hours, heating consumptions, and RH indoor values in the winter season for the three techniques W4, W5, and W6

Construction techniques/locality	Ancona November 1– April 15	Bolzano October 15– April 15	London September 15– June 15
Hours of overcooling PMV model			
% Hours on total			
Masonry	147	803	279
	3.7%	18.3%	4.2%
Wood-cement	142	793	274
	3.6%	18.1%	4.2%
Wood	195	809	306
	4.9%	18.4%	4.7%
Heating consumptions intermittent operation, 2 sections (5:00 a.m.−11:00 a.m., 3:00 p.m.−11:00 p.m.); kWh prim./m²year			
Masonry	19.43	41.82	38.26
Wood-cement	19.37	41.67	38.09
Wood	19.54	41.89	38.42
Average seasonal RH value (%)			
Masonry	58.5	54.7	62.9
Wood-cement	58.4	54.6	62.8
Wood	58.7	54.9	63.0

Source: F. Stazi, E. Tomassoni, C. Bonfigli, C. Di Perna, Energy, comfort and environmental assessment of different building envelope techniques in a Mediterranean climate with a hot dry summer, Appl. Energy 134 (December 1, 2014) 176−196, ISSN 0306-2619.

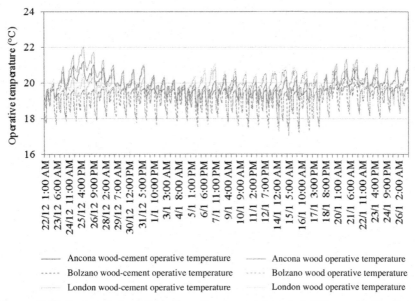

Ancona wood-cement operative temperature ——— Ancona wood operative temperature
····· Bolzano wood-cement operative temperature ····· Bolzano wood operative temperature
····· London wood-cement operative temperature London wood operative temperature

Figure 2.32 Numerical simulations. Operative temperatures in buildings with the three techniques in different climates in a typical winter week. *From Energy, comfort and environmental assessment of different building envelope techniques in a Mediterranean climate with a hot dry summer, Appl. Energy 134 (December 1, 2014) 176−196, ISSN 0306-2619.*

2.7 DESIGN PATTERNS

The relevant parameters to provide the behavior of a building envelope are its thermal transmittance and decrement factor for the winter and the internal areal heat capacity and the decrement factor for the summer. In hyperinsulated highly energy efficient envelopes the adoption of low values of the decrement factor combined with high values of internal areal heat capacity designate the best wall configuration when the aim is to maximize the comfort levels throughout the year. However, there is a lower limit for the f value around 0.04 under which overheating phenomena could occur.

Internal gains. If the apartment has high internal gains for the presence of high glazed surfaces or the absence of shadings/overhangs, it will behave as core-dominated environment. Differently it is skin dominated. In the former type, the adoption of high internal masses (high κ_1) is the priority. In the latter type, the best solutions are those with a very low decrement factor.

Window openings. The windows opening strongly reduces the difference between technologies. The study of surface temperatures at the ground floor of a single family house showed that the internal surface temperatures are very slightly influenced by windows opening or closing for both the massive and the lightweight walls because of the great incidence of the ground floor heat dispersions.

User behavior. The occupants' behavior has a great incidence on the efficacy of a retrofit intervention. For example, the scarce use of the heating plant for a social house building almost nullifies the benefices of the adoption of a very efficient energy saving envelope.

Heat losing elements. Another particular feature of the enclosed environment that has a great incidence on the maximization of the mass benefices in hyperinsulated envelope is the absence of heat losing elements, such as uninsulated horizontal slabs laying adjacent to the ground floor. The great rate of heat dispersion through this fabrics makes the vertical envelope optimization uneffective.

Climates. The comparison between summer and winter discomfort hours shows that the internal mass (high κ_1 value) is more important in hot and dry periods (summer and intermediate seasons) and for hot climate zones rather than winter periods or cold climates.

For the extreme climates (very hot as Cairo, or cold as Bolzano) the too high or too low temperatures do not allow the mass to completely dissipate the stored heat if overheated or to warm itself if overcooled. In such climates solutions with high global mass (very low f value) behaving as thermal barriers should be preferred.

REFERENCES

[1] P. Burberry, Environment and Services, Mitchell's Building Series, first ed, Longman, UK, 1997.

[2] N.K. Bansal, G. Hauser, G. Minke, Passive Building Design, A Handbook of Natural Climatic Control, Elsevier Science B.V., Netherlands, 1994.

[3] EN ISO 13786:2007. Thermal Performance of Building Components—Dynamic Thermal Characteristics—Calculation Methods.

[4] www.mygreenbuildings.org.

[5] F. Fantozzi, P. Galbiati, F. Leccese, G. Salvadori, M. Rocca, Thermal analysis of the building envelope of lightweight temporary housing, J. Phys. 547 (2014) 012011.

[6] M.R. Hall, Materials for Energy Efficiency and Thermal Comfort in Buildings. Woodhead Publishing Series in Energy: Number 14, 2010.

[7] F. Stazi, E. Tomassoni, C. Bonfigli, C. Di Perna, Energy, comfort and environmental assessment of different building envelope techniques in a Mediterranean climate with a hot dry summer, Appl. Energy 134 (2014) 176−196.

[8] E. Di Giuseppe, Nearly Zero Energy Buildings and Proliferation of Microorganisms, A Current Issue for Highly Insulated and Airtight Building Envelopes, Springer International Publishing, Cham, 2013.

[9] C. Di Perna, F. Stazi, A. Ursini Casalena, M. D'Orazio, Influence of the internal inertia of the building envelope on summertime comfort in buildings with high internal heat loads, Energy Build. 43 (2011) 200−206.

[10] C. Baglivo, P.M. Congedo, A. Fazio, Multi-criteria optimization analysis of external walls according to ITACA protocol for zero energy buildings in the Mediterranean climate, Build. Environ. 82 (2014) 467−480.

[11] F. Stazi, A. Vegliò, C. Di Perna, P. Munafò, Experimental comparison between 3 different traditional wall constructions and dynamic simulations to identify optimal thermal insulation strategies, Energy Build. 60 (2013) 429−441.

[12] V. Corrado, I. Ballarini, S. Paduos, Assessment of cost-optimal energy performance requirements for the Italian residential building stock, Energy Proc. 45 (2014) 443−452.

[13] E. De Angelis, G. Pansa, E. Serra, Research of economic sustainability of different energy refurbishment strategies for an apartment block building, Energy Proc. 48 (2014) 1449−1458.

[14] A.L. Pisello, F. Asdrubali, Human-based energy retrofits in residential buildings: a cost-effective alternative to traditional physical strategies, Appl. Energy 133 (2014) 224−235.

Retrofit of Existing Envelopes

3.1 INTRODUCTION

The 2012/27/UE Directive has established a common framework for the promotion of the energy efficiency in Europe and has underlined the importance of the renovation of the existing building stock because it represents the biggest potential for energy savings also highlighting the centrality of the residential sector. However, the energy saving standards, born in cold climates, were mostly focused on winter heating consumptions and have led even in warm Mediterranean climates, to the introduction of thick insulation layers on all existing envelopes, regardless their type. The introduction of thick insulations on envelopes characterized by unequal ways to interact with the external environment has a different effect depending on the adopted configurations and materials.

This chapter introduces the three main existing envelope typologies and explores their different behavior through experimental measures and analytical extensions. It compares retrofit alternatives and underlines the unequal effect of superinsulation on the three typologies. Finally, it identifies the most suitable retrofit solutions to maximize the sometimes contrasting aspects of energy saving, comfort, and global convenience. For envelopes with existing insulations, it reports the survey of the actual conservation state of the insulation material after several years of service life and suggestions on the most convenient interventions.

3.2 CAPACITY (C), STRATIFICATION (S), AND HIGH RESISTANCE (HR): THREE DIFFERENT WAYS AGAINST CLIMATE

Traditionally in temperate climates, three different typologies have been adopted in existing envelopes [1,2] for the protection against outdoor climate conditions (Fig. 3.1):

Thermal Inertia in Energy Efficient Building Envelopes.
DOI: http://dx.doi.org/10.1016/B978-0-12-813970-7.00003-0

Capacity (C series)   Stratified (S series)   High resistence (HR series)

Figure 3.1 Three types of existing building envelopes.

- *Capacity envelopes*: typical of traditional buildings in which envelopes with high thermal mass act as heat (and moisture) storage, leading to temperature, and humidity mitigation. Many authors [1−11] have demonstrated the efficiency of this system even compared with modern envelopes [12];
- *Stratified envelopes*: typical of buildings from the 1960s onward, in which perforated bricks and internal air cavities slow down the heat flux because each layer provides a slight resistance to the heat transfer [13−15];
- *High resistance envelopes*: adopted in buildings from 1980s until today, in which the protection against the environment consists in adopting thermal performing insulation materials that act as thermal barriers against heat loss. They are often made with multilayered walls or insulated dense prefabricated concrete panels that greatly reduce the rate of dissipation of internal humidity [1]. In several cases, they present 20 or 30 years old insulation layers with unknown actual performance. Several authors have shown the efficacy of this strategy, but the optimal insulation layer position could depend on the climate and the specific use of the apartment, intermittently or continuously heated [11,16−18].

In the past century there has been a transition of the building envelope from a traditional massive fabric, which adapt dynamically to the climate without the use of heating systems (as a sort of building membrane), to a lightweight and hyperinsulated assembly acting as a thermal barrier to the outdoor climate conditions. The recent energy saving regulations have adopted the latter strategy of highly thermal resistant enclosure thus imposing increasing thickness of thermal insulations also on existing buildings, regardless the specific envelope type and climate.

However, the introduction of high thermal insulations on the three identified envelope typologies has a very different impact on the final energy performance and comfort levels. If applied, e.g., on existing high masses (capacity wall) it will strongly compromise their dynamic behavior by impeding the heat release toward outside. Differently, it will result in a minor effect on lightweight stratified envelopes. Moreover, if existing insulations with long service life are present, different problems should be addressed. The presence of any preexisting insulating layer subject to high moisture levels, especially if adjacent to the internal rooms, poses the problem to quantify its actual performance and to decide whether remove or maintain it.

To contribute to the debate regarding this open questions the present chapter reports the results of simultaneous experimental investigations on three existing envelopes, representative of the three identified envelope typologies, underlining their very different behavior. Optimal retrofit solutions are drawn for each of the three envelopes through comfort, energy, and environmental analysis also considering the behavior at the varying of some parameters: the dimension of the existing mass, the type and position of thermal insulation, the heating operation profile. The variation was done not only on the vertical walls but also on the horizontal external floors (roof and ground floor slabs) by adopting the same technology (equal insulation-mass position and insulation level) selected for the vertical walls. In fact, the presence of high heat losses would have strongly reduced the differences on the performance of the compared envelopes.

3.3 SIMULTANEOUS MEASURE OF THE THREE DIFFERENT EXISTING ENVELOPE TYPOLOGIES

3.3.1 Problem description

The present section shows through simultaneous experimentations the very different behavior of three envelope types in the same environmental conditions.

Preliminary typological, constructive, and statistical analyses allowed selecting three reference buildings representative for the three identified traditional wall constructions and characterized by similar features regarding glazed percentage, geometry, orientation, type of occupancy. Summer and winter simultaneous monitoring were carried out on three social hosing buildings: building B4 (wall C1), building B6 (wall S1), and the same building type B6 after a retrofit intervention (wall HR1). The description of buildings features is reported in Appendix A, while the thermal parameters are detailed in Table 3.1.

Table 3.1 Thermal parameters for three envelopes representative of as many existing envelope typologies

Three strategies	C1	S1	HR1
Description	Solid brick masonry	Unfilled brick-block cavity wall	Unfilled brick cavity wall with existing insulation
Wall thickness (cm)	50	34	40
Thermal transm. U (W/(m^2 K))	1.34	1.2	0.4
Periodic transm. Y_{ie} (W/(m^2 K))	0.076	0.5	0.11
Internal capacity κ_1 (kJ/(m^2 K))	60	50	44
f (−)	0.06	0.41	0.26

3.3.2 Summer behavior

Referring to the incidence of dynamic parameters on the summer behavior as identified in Sections 2.2.4–2.2.6 and to proposed limits (Section 2.2.9), a good summer performance is expected for the three walls, since low f value are present (but not excessively low) and all the walls have high κ_1 values. The worst behavior is expected for wall S1 for its lower ability to attenuate the incoming heat wave in response to sol-air fluctuations (since it has a too high f value), while the best will be C1, for its favorable combination of dynamic parameters.

Fig. 3.2 on the left shows the simultaneous comparison between the summertime experimental results for the solid brick masonry (C1) and the unfilled cavity wall (S1). Fig. 3.2 on the right side reports the same building with wall S1 compared with the unfilled cavity wall with existing 5 cm expanded polystyrene coating (HR1). Being the outdoor environmental conditions very similar for the two periods, it is also possible to make horizontal comparisons.

The external surface temperature of the solid brick wall C1 is always lower than that of the cavity wall S1 since the former acts as a heat storage system. The external surface temperature of the insulated wall HR1 is even higher than the cavity wall S1 ($+3°C$) in the hottest part of the day and shows instead lower values (of about $4°C$) during the night. The overheating of the external surface of the insulation coating during the hottest hours of the day, with temperatures of up to $40°C$, should not be underestimated when evaluating the durability of this solution.

Regarding the internal air temperatures the solid brick wall (C1) and the insulated wall (HR1) provide similar values with a constant trend since, having a high attenuating attitude of the crossing thermal wave, they are slightly influenced by the outdoor conditions. On the contrary, the internal air temperature of the building with the cavity wall (S1) has greater variations.

An analysis of the internal surface temperatures shows similar values for walls C1 and HR1, while the wall S1 has values up to $4–5°C$ higher with a more variable trend.

The heat flux in solid brick wall (C1) is almost always outgoing (in summer an outgoing heat flux is conventionally considered negative) with values down to -6 W/m^2, demonstrating how, throughout the day, the wall takes the heat away from the indoor environment. Instead, the flux in the building with the cavity wall (S1) is much more variable, being

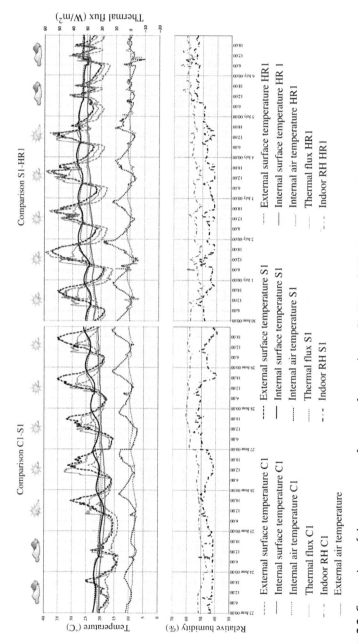

Figure 3.2 Comparison of the summer performance for envelopes C1, S1, and HR1 (experimental measures). *From F. Stazi, A. Vegliò, C. Di Perna, P. Munafò, Experimental comparison between 3 different traditional wall constructions and dynamic simulations to identify optimal thermal insulation strategies, Energy Build. 60 (May 2013), 429–441, ISSN 0378-7788.*

positive or negative according to the time of day, since it is strongly influenced by the outdoor temperatures. In the sunny period, for most of the day the external surface temperature has higher values than the internal surface temperature, and therefore the flux is incoming, with peaks of up to 11 W/m^2. However, in the morning the flux is generally outgoing with values going down to -6 W/m^2. The insulated wall (HR1) has a reduced crossing flux, with maximum incoming values of 2 W/m^2 and a maximum outgoing one of -2 W/m^2.

Analyzing the relative humidity of the room with solid brick wall (C1) and that with insulated wall (HR1), the values are on average around 50%–60% with an almost constant trend. Differently the relative humidity in the building with the cavity wall (S1) is between 40% and 50%, with an average difference of 10% compared with the other two and with greater fluctuations. This difference is due to higher indoor temperatures recorded in such building that even with similar water vapor content determines lower RH values (for major vapor "solubility" at high temperatures).

3.3.3 Winter behavior

Referring to the incidence of dynamic parameter identified in Section 2.2.8 and to the proposed limits (Section 2.29), in which it is highlighted that the winter performance is mainly influenced by both U and f values, the best solution is expected to be wall HR1 for its lowest U value. Between C1 and S1 (with similar U values), the former is expected to be a preferable solution for its very lower f value.

Fig. 3.3 on the left shows the comparison between the winter results for the solid brick masonry (C1) and the unfilled cavity wall (S1). Fig. 3.3 on the right compares S1 with the unfilled cavity wall with existing 5 cm expanded polystyrene coating (HR1). For the similarity of the outdoor conditions on the two periods, it is also possible to make horizontal comparisons between the three techniques.

The external surface temperature of the solid brick wall (C1) is between 0°C and 11°C, while that of the cavity wall (S1) is between -3°C and 18°C, with greater fluctuations. The latter wall have a trend similar to those recorded in HR1 except in the middle of the day (1.00 p.m.) when its temperature is as much as 15°C lower than the insulated wall, that is overheated by direct sun radiation.

The internal air temperature of the building with solid brick wall (C1) is between 8°C and 15°C, while in the building with cavity wall

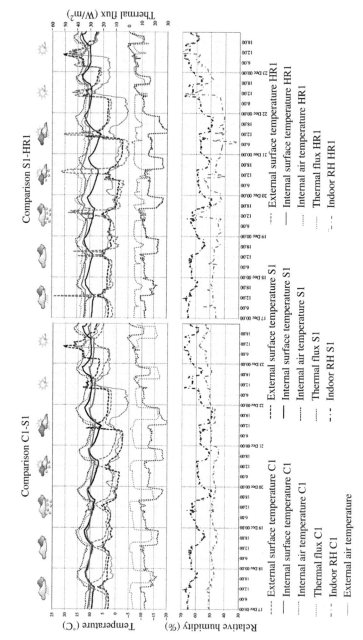

Figure 3.3 Comparison of the winter performance for envelopes C1, S1, and HR1 (experimental measures). Note that the secondary y-axis is overturned since in winter an outgoing heat flux is conventionally considered positive. This choice is to favor a transversal comparison with the image relative to the summer period. *From F. Stazi, A. Veglio, C. Di Perna, P. Munafò, Experimental comparison between 3 different traditional wall constructions and dynamic simulations to identify optimal thermal insulation strategies, Energy Build. 60 (May 2013), 429–441, ISSN 0378-7788.*

(S1) it ranges between 9°C and 16°C; in the building with insulated wall (HR1) it is between 12°C and 17°C. It should be underlined that the internal air temperatures never go over 17°C because in all buildings there is a very limited use of the heating system.

The internal surface temperature of the solid brick wall (C1) is between 8°C and 12°C, while with the cavity wall (S1) it has greater variation ranging between 6°C and 13°C. The surface temperature of the insulated wall (HR1) ranges between 10°C and 14°C, with a difference of up to 4°C.

By analyzing the heat flux it is possible to note that for all the walls the flux is always outgoing (refer to secondary y-axis) throughout the period, and it has highest values for the cavity wall (S1). The heat flux through the solid brick wall (C1) has peaks of 18 W/m^2, while the flux through the cavity wall (S1) reaches 20 W/m^2. On the contrary, the insulated wall (HR1) has the lowest outgoing flux (with peaks of 14 W/m^2) for the presence of the external insulation layer, even if it is only 5 cm thick.

The relative humidity of both buildings with solid brick walls (C1) and with insulated walls (HR1) is between 45% and 55% and has a constant trend. Instead, the relative humidity of the building with cavity wall (S1) is on average between 55% and 65%, with a difference of 10% compared with the solid brick masonry building (C1).

3.4 RETROFIT OF CAPACITIVE LOAD-BEARING WALLS, EXPLORING DIFFERENT C TYPES

3.4.1 Problem description

For massive load-bearing envelopes (C types) mixed solutions should be identified which, while using considerable thermal resistance for the winter, are able to exploit the dynamic behavior of the mass placed adjacent to the room in the hot periods, through an interaction with the outdoor and indoor environments that varies according to needs. The thicker is the existing mass, the greater is the importance of adopting solutions that guarantee its dynamic behavior.

The present section deepens this aspect.

Consider the walls shown in Table 3.2. The existing high masses have a very pronounced damping attitude (down to 0.06 for the thickest wall)

Table 3.2 Thermal parameters for three capacitive envelopes with different mass and for the retrofit interventions

Solid masonry four-wythe

	C1	C1ext	C1int	C1vent
Ground floor building B.4 Experimental data in Section 3.3				
Description	Solid brick	External insul. 9 cm	Internal insul. 9 cm	Ventilated layer[a]
Wall thickness (cm)	50	60.5	60.5	69.5
Thermal transm. U (W/(m^2 K))	1.34	0.3	0.3	0.29
Periodic transm. Y_{12} (W/(m^2 K))	0.076	0.003	0.006	0.003
Int. areal h. capacity κ_1 (kJ/(m^2 K))	60	60	21	60
f (−)	0.06	0.01 Too low	0.02 Too low	0.01 (winter)
U ground floor slab (W/(m^2 K))	1.86	0.35	0.35	0.35

Solid masonry three-wythe

	C2	C2ext	C2int	C2vent
Ground floor building B.3 Experimental data in Section 2.5				
Description	Solid brick	External insul. 12 cm	Internal insul. 12 cm	Ventilated layer[a]
Wall thickness (cm)	45	57	57	65
Thermal transm. U (W/(m^2 K))	1.35	0.25	0.25	0.24
Periodic transm. Y_{12} (W/(m^2 K))	0.18	0.01	0.01	0.006
Int. areal h. capacity κ_1 (kJ/(m^2 K))	66	63	28	63
f (−)	0.14	0.029 Low	0.05	0.026 (winter)
U ground floor slab (W/(m^2 K))	1.44	0.3	0.3	0.3

Semi-solid bricks	C3	C3ext	C3int	C3vent
First floor building B.3 Experimental data in Section 2.5				
Description	Light masonry	External insul. 12 cm	Internal insul. 12 cm	Ventilated layer[a]
Wall thickness (cm)	28	40	40	48
Thermal transm. (W/(m^2 K))	1.11	0.24	0.24	0.23
Periodic transm. Y_{12} (W/(m^2 K))	0.42	0.01	0.037	0.014
Int. areal h. capacity κ_1 (kJ/(m^2 K))	58	53	28	57
f (−)	0.37	0.09	0.16	0.06
U roof slab (W/(m^2 K))	1.91	0.24	0.24	0.23[b]

[a]Ventilated air gap, closed in winter and open in summer.
[b]This case regards the introduction of a ventilated layer even in the roof slab.

and the introduction of the insulation layer reduces even more the decrement factor value thus configuring thermal barriers. The higher is the existing mass thickness (going e.g., from a three-wythe C2 to a four-wythe solid brick masonry C1) the higher will be the thermal barrier effect after having applied the insulation layer. In this latter case, the best solution is the adoption of dynamic insulations that could be deactivated during the hottest summer periods.

The following results derive from parametric variations on virtual models calibrated through experimental measures (in buildings B3 and B4, see Appendix A).

3.4.2 Solid masonry four-wythe C1

The study of the behavior of the four-wythe masonry building (C1) at the varying of insulation measure demonstrates the importance of adopting dynamic solutions to avoid the summer overheating. Instead, the different insulation measures have similar results in wintertime.

Figs. 3.4 and 3.5 show, respectively, the surface and operative temperatures of the different retrofit interventions on an apartment at the ground floor of building B.4.

During the coldest week of the winter the as built wall W1 has very low internal surface temperatures with maximum peaks of 17°C and minimum peaks of 14.5°C, values far below the thermal comfort standards. The retrofitted walls show similar trends with flatter plots and values up to 3°C higher than the as built wall. Wall W1vent has the best behavior for its highest thermal resistance and lowest f value, given by an additional layer of air in steady state conditions.

In summer, the internal surface temperature of the "as built" four-wythe solid brick masonry (C1) ranges between 23°C and 26°C with a constant trend and small fluctuations. The insertion of 9 cm of expanded polystyrene in the summer season has a negative effect, as it leads to an increase up to 2°C.

The ventilated system values stand lower than the other intervention curves. The benefits of adopting a vented system respect to a traditional external insulation could be quantified in a reduction of 2–3°C compared with a traditional coating (C1ext), in typical summer days, thus avoiding the previously discussed problem of summer overheating which occurs in the latter case. This kind of intervention seems to be of high importance in very massive existing envelopes.

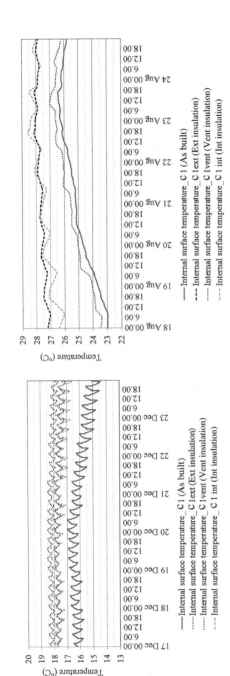

Figure 3.4 Winter and summer surface temperatures for envelopes C1, C1ext, C1int, and C1vent (simulations on the model calibrated on experimental measures). *From F. Stazi, A. Vegliò, C. Di Perna, P. Munafò, Experimental comparison between 3 different traditional wall constructions and dynamic simulations to identify optimal thermal insulation strategies, Energy Build. 60 (May 2013), 429–441, ISSN 0378-7788.*

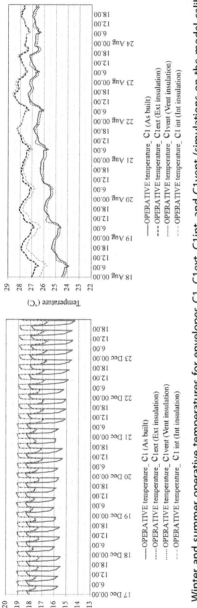

Figure 3.5 Winter and summer operative temperatures for envelopes C1, C1ext, C1int, and C1vent (simulations on the model calibrated on experimental measures).

3.4.3 Solid masonry three-wythe C2

The study of the behavior of the three-wythe masonry building (C2) at the varying of insulation measure demonstrates once again the best performance of the vented solution, even if with less appreciable benefices respect to the previous very thick masonry wall (C1). The experimental data of the building with wall C2 (at the ground floor) are reported in Section 2.5.

Figs. 3.6 and 3.7 show, respectively, the surface and operative temperatures obtained for the retrofit of the whole envelope (floor and wall) at the ground floor level of building B.3. The graphs confirm the previous results, since all retrofit interventions increase the internal surface temperatures respect to the initial "as built" situation. The ventilated system values stand lower than the other intervention curves. The benefits of adopting a vented system respect to a traditional external insulation in typical summer days could be quantified in a reduction of 1−1.5°C in the internal surface temperatures and 0.5°C in the operative temperature, while lower benefices are seen in hot days with external extremely high temperatures.

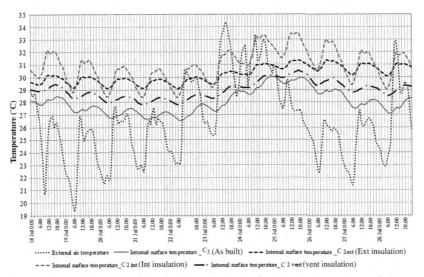

Figure 3.6 Summer internal surface temperatures for C2, C2ext, C2int, and C2vent (simulations on the model calibrated on experimental measures). *From F. Stazi, C. Bonfigli, E. Tomassoni, C. Di Perna, P. Munafò, The effect of high thermal insulation on high thermal mass: is the dynamic behaviour of traditional envelopes in Mediterranean climates still possible?, Energy Build. 88 (February 1, 2015) 367−383, ISSN 0378-7788.*

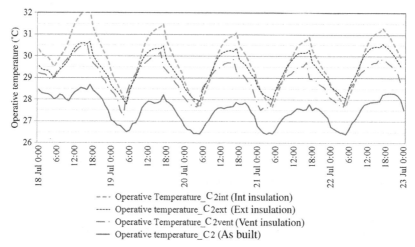

Figure 3.7 Summer operative temperatures for envelopes C2, C2ext, C2int, and C2vent (simulations on the model calibrated on experimental measures).

In the intermediate seasons (Fig. 3.8) the comparison between the retrofit techniques with high internal mass (C2ext) and low internal mass (C2int) allows to identify the "trend inversion" as highlighted in Section 2.5.1. The vented solution curve was not reported since, having in the selected period the vents closed, the values were almost coincident with the traditional external insulation solution.

During the spring season (Fig. 3.8A) different phases could be identified. A first phase (March 1−5) in which both walls are still affected by the typical winter behavior (heated room) since the heating system was recently turned off (on March 1 for this simulation). The operative temperatures have the same minimum values, while the room with internal insulation (low capacity) is characterized by greater maximum values (1.5−2°C). A second phase (March 5−17) when the room with lightweight envelope undergoes a sudden lowering in operative temperature values because the heating effect is ended and the outside temperatures are still low. In spring and for the whole summer, the room with lightweight wall presents peaks of overheating. Instead at the end of the hot period (September−October, Fig. 3.8B) the external air temperature drops and causes especially for this solution the internal gradual reduction of the operative temperatures (up to 3°C less than the massive wall), with an unfavorable behavior for the approaching of the cold season.

The analysis of discomfort hours (Table 3.3) shows that, as resulted from the summer monitoring (Fig. 2.23), the "as built" condition is

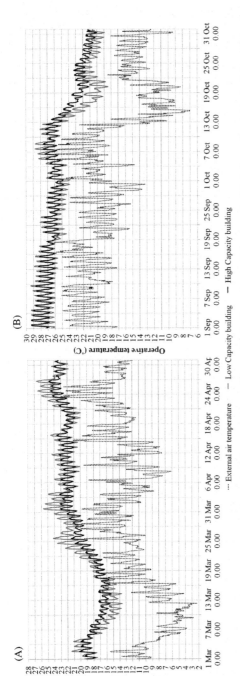

Figure 3.8 Intermediate seasons performance for envelopes C2ext (High Capacity building) and C2int (Low Capacity building) (simulations on the model calibrated on experimental measures). (A) March—April and (B) September—October. *From F. Stazi, C. Bonfigli, E. Tomassoni, C. Di Perna, P. Munafò, The effect of high thermal insulation on high thermal mass: is the dynamic behaviour of traditional envelopes in Mediterranean climates still possible?, Energy Build. 88 (February 1, 2015) 367—383, ISSN 0378-7788.*

Table 3.3 Summer discomfort hours for different retrofit interventions on the three-wythe masonry apartment (ground floor, simulations on the model calibrated on experimental measures)

		Discomfort hours (h)[a]	
		Overheating	Overcooling
	As built	0	378
	Windows optimization (U glass-frame = 2 W/m^2 K)	0	246
	Ground floor slab insulation external side/ internal side (U = 0.3 W/m^2 K)	130/183	0/4
	G. slab ext + Wall C2 ext	193	0
	G. slab ext + Wall C2 vent	72	0
	G. slab int + Wall C2 int	542 (329)[b]	0

[a]Simulated values with a continuously natural vented environment.
[b]Addition of a clay panel on the inner wall side (thermal characteristics: λ = 0.047 W/(m K); c = 1000 J/(kg K); ρ = 1300 kg/m^3).

characterized by overcooling (about 378 hours) at the ground floor for the heat dispersion toward the ground. The windows thermal performance optimization and the simultaneous increasing of the glazed surface (as requested by the current healthy regulations) slightly reduce the ground floor overcooling (from 378 to 246 hours). The insulation of ground floor causes (regardless the insulation position) a considerable reduction of the overcooling discomfort hours (with values down to 0−4 hours), but increases the overheating with slightly preferable comfort levels by leaving the slab mass on the innermost side. This confirms the results obtained in Section 2.5.4 (adiabatic layer). The previous scenarios were improved with the subsequent insulation of the vertical walls (by combining the same technologies for both walls and slabs), with the alternative presented in Table 3.2. The results demonstrate that the improvement or worsening of the comfort conditions strictly depend on the adopted insulation position (exterior, ventilated, or interior). The high capacity-building envelope characterized by external insulation layer worsens comfort levels (compared to the previous scenarios) by increasing the overheating discomfort hours (from 130 to 193 hours). The insulation material applied on the inner side causes very high discomfort level due to overheating almost tripling the discomfort hours (from 183 to 542 hours). Differently from the other two solutions, the ventilated insulation system ensures a good indoor thermal comfort by reducing the discomfort hours down to 72. It takes advantage by cold nocturnal air that in this wall flows adjacent to the inner mass with a cooling effect. The introduction of a massive clay panel as internal finishing in the low massive wall (C2int) determines a reduction of the discomfort hours. The value decreases down to 329 hours from an initial value of 542 bringing comfort levels more close to the High Capacity building with external insulated walls (C2ext).

3.4.4 Semisolid masonry C3

For the semisolid brick masonry, the experimental data (building B.3 with wall C3 at the first floor) are reported in Section 2.5.

In this technique, the adoption of a naturally ventilated insulation layer in summer is preferable also with respect to the existing (not insulated) wall (Fig. 3.9). The low global inertia (high f value) of the wall makes an insulation intervention more important to block the incoming heat (and the consequent reduction of the surface temperature

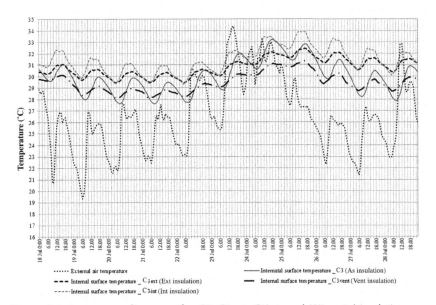

Figure 3.9 Summer performance for C3, C3ext, C3int, and W3vent (simulations on the model calibrated on experimental measures). *From F. Stazi, C. Bonfigli, E. Tomassoni, C. Di Perna, P. Munafò, The effect of high thermal insulation on high thermal mass: is the dynamic behaviour of traditional envelopes in Mediterranean climates still possible?, Energy Build. 88 (February 1, 2015) 367–383, ISSN 0378-7788.*

fluctuations). This effect is even more evident in the extremely hot days (July 22–25), and in general in extreme climates, in which the primarily required building envelope performance is to exclude the entering heat flow.

The analysis of discomfort hours (Table 3.4) confirms the presence of overheating (128 hours) as resulted from the summer monitoring of the "as built" condition (see Fig. 2.23 first floor level with closed windows). The thermal performance optimization of the windows and the simultaneous increasing of the glazed surface increase the overheating hours up to 144. This result differs to what occurred at the ground floor (see Table 3.3) in which the enlargement of the windows was beneficial for the presence of a strongly overcooled environment. The insulation of the roof slab causes (regardless the insulation position) a considerable reduction of the overcooling discomfort hours (with values down to 4–6), but almost doubles the overheated hours with preferable comfort levels if adopting the external insulation. The previous scenarios were combined with the insulation of the vertical walls. The results demonstrate that the improvement or worsening of the comfort conditions strictly depend on the adopted insulation position (exterior, ventilated, or interior). The high-capacity building envelope characterized by external insulation layer worsens comfort levels (compared to the previous scenario) by increasing the overheating discomfort hours from 227 to 356. The insulation material applied on the inner side causes very high overheating almost tripling

Table 3.4 Winter discomfort hours for different retrofit interventions on the semisolid brick masonry apartment (first floor)

		Discomfort hours (h)[a]	
		Overheating	Overcooling
	As built	128	83
	Windows optimization ($U = 2$ W/m^2 K)	144	75
	Roof slab insulation (ext) ($U = 0.24$ W/ m^2 K)	227	4
	Roof slab insulation (int) ($U = 0.24$ W/ m^2 K)	281	6
	Roof slab ext + Wall C3 ext	356	0
	Roof slab vent + Wall C3 vent	93	0
	Roof slab vent + Wall C3 int	655 (463)[b]	0

[a]Simulated values with a continuously natural vented environment.
[b]Addition of a clay panel on the inner wall side.

the discomfort hours from 281 to 655. Differently from the other two solutions, the ventilated insulation system ensures a clear improvement of the indoor thermal comfort conditions by reducing the discomfort hours down to 93. This system is the only insulation configuration, which enhances the comfort conditions. It takes advantage by cold nocturnal air that in this wall and in the roof slab flows adjacent to the inner mass with a cooling effect. The introduction of a massive clay panel as internal finishing in the low massive wall determines a reduction of discomfort hours down to 463 hours from an initial value of 655, bringing values more close to the High Capacity building (C3ext).

3.4.5 Optimal retrofit intervention from a global cost evaluation

What are the consumption of the entire building (building B.3 in Appendix A with C2 walls at the ground floor and C3 walls at the first floor) before and after the interventions? What is the most convenient solution from a global cost evaluation? The following data respond to these questions.

Table 3.5 shows the consumptions evaluation in terms of primary energy for both heating and cooling demand. Regarding winter consumption, the glazed surface enlargement and its thermal optimization lead to a heating consumption reduction for both continuous and intermittent operation (about 8% in the first case and 7% in the second). This is due to the increase of solar gains and the simultaneous improvement of the thermal performance of the existing glazed surface. For the same reason, however, this intervention causes a slight increase in summer consumption. The insulation of the horizontal floors (ground floor and roof) reduces the winter consumptions down to about 106 kWh/m^2 year for a continuous system operation and to about 81 kWh/m^2 year for intermittent use, regardless of the reciprocal position (external or internal) between insulation layer and mass. The summer consumption is even more increased. The subsequent optimization of the walls results in a significant reduction in primary energy winter consumption by placing the insulation layer on the outer side, with almost similar performance between the traditional insulating system and the vented one. The latter solution is slightly better because of the higher thermal resistance due to the addition of an air cavity (not vented in winter period). Compared to these two interventions the low inertia retrofit with insulation placed on the internal side has higher consumptions (around 41 kWh/m^2 year for

Table 3.5 Winter and summer consumptions for retrofit measures on the building with walls C2 at the ground floor and C3 at the first floor

Consumptions (kWh/m² year)	Wall Ground floor	Wall First floor	Slab Ground floor	Slab Roof	Continuous heating	Intermittent heating	Summer cooling
1	As built (C2)	As built (C3)	As built	As built	138.62	102.88	14.18
2	Optimized windows	Optimized windows	As built	As built	128.71	95.95	14.57
3	Optimized windows	Optimized windows	External insulation[a]	External insulation	105.93	81.07	25.15
4	C2 + optimized windows	C3 + optimized windows	Internal insulation[a]	Internal insulation	105.47	80.82	25.77
5	C2ext	C3ext	External insulation	External insulation	38.31	30.93	27.86
6	C2vent	C3vent	External insulation	External insulation	37.67	30.35	22.19
7	C2int	C3int	Internal insulation	Internal insulation	41.22	33.34	30.69
8	C2int (clay)[b]	C3int (clay)	Internal insulation	Internal insulation	40.33	32.61	28.18
9	C2ext	C3ext	External insulation	Ext. insulation attic slab removal	34.59	28.21	29.19
10	C2vent	C3vent	External insulation	Ext. insulation attic slab removal	33.92	27.59	24.32
11	C2vent	C3vent	External insulation	Vent. insulation[c] attic slab removal	33.69	27.41	17.30

[a]External insulation is positioned leaving the mass of the concrete horizontal floor on the inner side; internal Insulation is adjacent to the internal environment.
[b]Addition of a clay panel on the inner wall side.
[c]Ventilated insulation (see Fig. 3.12).

continuous ignition and 33 kWh/m² year for intermittent ignition). In this case, the adoption of an internal massive finishing slightly reduces winter and summer consumptions.

The analysis of summer consumptions shows that the insulation interventions (insulation of windows, roof, ground floor, and walls) worsen the "as built" condition in all studied configurations. The benefit of adopting a ventilated solution is noticeable, since it allows a dynamic behavior of the inner mass through the ventilation of the internal gap. Moreover, there are 13 kWh/m² year difference between this preferable vented solution and the worst Low Capacity solution (C2int + C3int) because the insulating layer placed on the inner side causes overheating phenomena.

The totally vented configuration (scenario number 11) allows a further reduction respect to the building where vented wall and traditionally insulated roof are instead adopted (solution number 10), reaching a minimum value of 17 kWh/m² year and saving up to 18% for winter heating and 43% for summer cooling respect to the worst case outcome.

The envelope optimization at the first floor has a different effect depending on the presence of an existing attic floor (solutions 5−8) or its removal (solutions 9−11). Moreover, the ventilation of the roof slab through the introduction of an air gap interposed between the external insulating material and the inner massive floor would produce appreciable benefices only in the latter case, i.e., without a separating attic space. The latter solution presents the lowest consumptions. A detailed explaining of this benefice is given in Fig. 3.10 that shows the operative temperatures for the presence or absence of the subdividing attic floor for a typical summer day. In a traditional intervention with external insulation laid adjacent on both walls and external floors (HC building, *light gray lines*), the removal of the attic floor causes a slight increase of the operative temperatures. The attic floor acts as a thermal buffer space between the internal environment and the roof structure. The adoption of a configuration characterized by ventilated insulation on walls and a traditionally insulated roof (HC vented building, *dark gray lines*) lowers the operative temperatures to about 1°C. Even in this case, the maintenance of the attic floor is preferable for the abovementioned reason. The solution with ventilated insulation both on walls and roof (total vented solution, *black dashed line*), that involves the attic floor removal, guarantees a reduction of 2°C in operative temperature compared to a traditional insulation system.

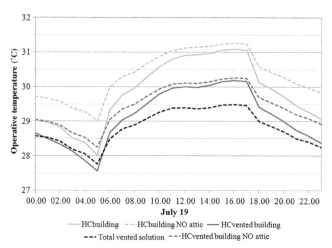

Figure 3.10 Incidence of different retrofit solutions with and without the attic floor slab.

The global cost assessment in relation to the overall energy performance was carried out for the different retrofit interventions. The graph in Fig. 3.11 shows the global cost for the scenario in which both heating and cooling system are included (the same internal comfort conditions between the various solutions are imposed), and the case in which the only heating system is used (excluding the final histograms quote), adopting a summer natural ventilation. In the latter case, the different retrofit solutions led to unequal summer comfort levels evaluated as the percentage of discomfort hours over the entire season (*dashed line*, referred to secondary *y*-axis). In the graph "HC" stands for High Capacity and "LC" for Low Capacity, depending on mass position within the envelope.

The interventions related to the single building element improvement (on the right side of the graph) regard the improvement of windows thermal performance and the internal or external insulation of the horizontal slabs (respectively LC floor and HC floor). Such measures are not convenient for the high cost related to winter heating. The introduction of internal insulation on both horizontal and vertical envelope (C2int + C3int) thus configuring a building characterized by low internal capacity (named "LC building" in Fig. 3.11) is not cost effective being characterized by higher global costs than the other solutions and by high summer discomfort levels. The preferable systems is that with external traditional insulation on the whole envelope leaving high capacity on inner side (namely "HC building") and that with ventilated insulation

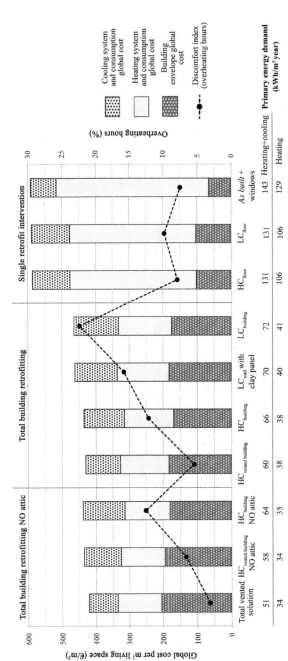

Figure 3.11 Global cost evaluation at the varying of the retrofit solution for the whole building. *From F. Stazi, C. Bonfigli, E. Tomassoni, C. Di Perna, P. Munafò, The effect of high thermal insulation on high thermal mass: is the dynamic behaviour of traditional envelopes in Mediterranean climates still possible?, Energy Build. 88 (February 1, 2015) 367–383, ISSN 0378-7788.*

(HC vented building). These solutions have a similar global cost but the latter guarantees lower discomfort levels if choosing to adopt a passive cooling strategy. The removal of the slab separating the first floor space from the attic determines an increase in the building envelope global cost for the additional cost of slab demolition. Nevertheless, the global cost of the building configuration with ventilated insulation applied both to walls and roof (named "total vented solution") is the lowest one because even

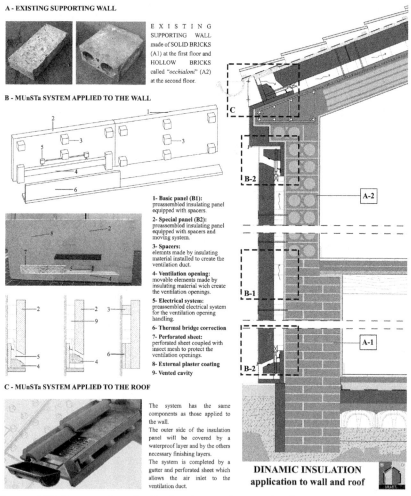

Figure 3.12 Construction details for the total vented solution. *From F. Stazi, C. Bonfigli, E. Tomassoni, C. Di Perna, P. Munafò, The effect of high thermal insulation on high thermal mass: is the dynamic behaviour of traditional envelopes in Mediterranean climates still possible?, Energy Build. 88 (February 1, 2015) 367–383, ISSN 0378-7788.*

if characterized by a greater initial investment it guarantees very lower summer consumption resulting to be cost effective by a global evaluation. Moreover if adopting a summer passive cooling (thus excluding the superior histogram quote), this last solution, despite characterized by a higher global cost, presents optimal indoor thermal comfort conditions.

An example of the constructive details for the total vented solution with attic slab removal is shown in Fig. 3.12.

3.5 RETROFIT OF STRATIFIED ENVELOPES, COMPARING ALTERNATIVES FOR RETROFIT OF S TYPES

3.5.1 Problem description

The cavity walls for the presence of an air gap could be retrofitted by adopting different materials (inorganic mineral materials or organic insulating materials) with unequal structures (mainly fibrous materials, foams, or

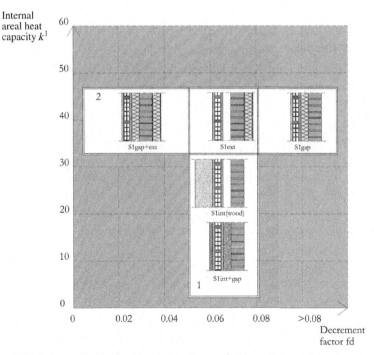

Figure 3.13 Scheme for different combinations of thermal dynamic parameters in existing stratified walls.

granulates) and in various positions (external, internal, interposed, or a combination [19]). So very different thermal conditions could be achieved.

Consider the existing stratified walls shown in Fig. 3.13 representing alternative retrofit interventions on the existing wall S1 (building B.6) experimentally surveyed in Section 3.3. The alternative solutions are detailed from a thermophysical point of view in the following subsections, by adopting the different materials proposed in Table 3.6. The thermal parameters for the initial as built solution S1 are reported in Table 3.1. The obtained walls after the retrofit interventions present decrement factor values too low only for $S1_{gap+ext}$, namely a stratified wall with insulation both external and in the gap, being equal to 0.032 (very near to the proposed limit of 0.04). In all the other walls as above demonstrated (see Section 2.2.9) the mass is still able to dissipate the heat toward the outside and vented insulation are not necessary. So the present section compares different traditional energy saving interventions and shows that for such envelopes high κ_1 values combined with low f values have a positive both summer and winter behavior. The wall named $S1_{gap+ext}$ shows (although negligible) overheating. The best solution(s) are identified.

3.5.2 Medium decrementing attitude ($0.04 < f < 0.08$) and variable inertia

Traditional cavity walls can be retrofitted achieving different levels of internal inertia, as graphically shown in Fig. 3.13, variations 1 and as detailed in Table 3.7. The highest κ_1 value will guarantee the best comfort condition throughout the year.

During the coldest week in the winter (Fig. 3.14 top graph), the behavior of the walls is strongly affected by the thermal transmittance values, which are the same in all three cases. As a result, the trends are quite similar. The cavity wall with external insulation ($S1_{ext}$) has a slightly flatter plot thanks to its greater internal inertia. In winter the operative temperature are not reported since very similar for the use of the heating plants.

In the hottest week of the summer (Fig. 3.14 bottom graph), the insertion of external expanded polystyrene (9 cm thick) allows to have a flatter plot, indicating the positive impact (especially in very hot days) of this insulating material in comparison with the as built initial condition, unlike the results obtained for the solid brick masonry (Section 3.4.2). In fact, in this case the f values reached after the insulation intervention are not too low thus allowing the mass cooling during the night.

Table 3.6 Thermal properties of the main materials of the external envelopes analyzed

Insulation materials (design values)	Polyurethane foam	Expanded polystyrene	Expanded clay	Mineralized wood fiber	Inflated granular mineral insulation	Perforated bricks	Solid bricks
Thermal conductivity λ (W/(m K))	0.033	0.036	0.09	0.076	0.052	0.25	0.78
Specific heat capacity c (J/(kg K))	1480	1480	940	1150	900	840	940
Density ρ (kg/m^3)	50	35	350	300	110	600	1700

Table 3.7 Thermal parameters for walls with medium global inertia (f value) and variable internal inertia

Medium global inertia solutions ($f < 0.08$)	S1$_{ext}$	S1$_{int\,(wood)}$	S1$_{int+gap}$
Description	External insulation layer (expanded polystyrene)	Mineralized wood insulation in the internal side	Internal expanded polystyrene and mineral insulation in the gap
	1. Plaster (1.5 cm)	1. Mineralized wood (19 cm)	1. Exp. polystyr. (4 cm)
	2. Perforated bricks (8 cm)	2. Plaster (1.5 cm)	2. Plaster (1.5 cm)
	3. Cem. mortar (1.5 cm)	3. Perf. bricks (8 cm)	3. Perf. bricks (8 cm)
	4. Air gap (9 cm)	4. Cem. mortar (1 cm)	4. Cem. mortar (1 cm)
	5. Solid bricks (12 cm)	5. Air gap (9 cm)	5. Mineral insul. (9 cm)
	6. Adhesive (0.5 cm)	6. Solid bricks (12 cm)	6. Solid bricks (12 cm)
	7. EPS (9 cm)		
	8. Plaster (0.5 cm)		
Thickness (cm)	42	50.5	35.5
U (W/(m^2 K))	0.29	0.29	0.29
Y_{12} (W/(m^2 K))	0.021	0.017	0.025
κ_1 (kJ/(m^2 K))	40 High	19 Medium	11 Low
f [−]	0.07 Medium	0.06 Medium	0.08 Medium

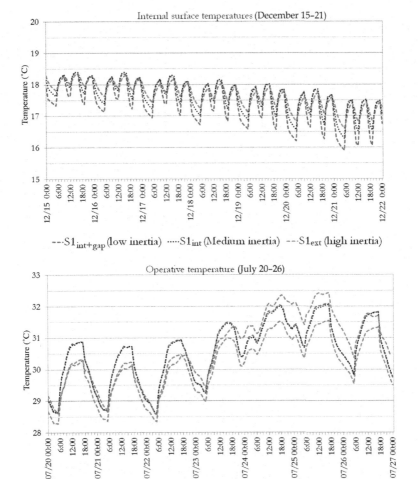

Figure 3.14 Medium *f* and variable inertia: winter and summer behavior of walls S1$_{int+gap}$, S1$_{int}$ (wood), and S1$_{ext}$ (simulations on the model calibrated on experimental measures).

3.5.3 High thermal inertia ($\kappa_1 > 40$) and variable decrementing attitude

Adopting the optimal value of the internal inertia (value of about 40 kJ/ (m^2 K)), the comparison with alternative walls (as graphically shown in Fig. 3.13 variations 2 and in Table 3.8), shows that the solutions with minimum decrement factor are preferable and only negligible overheating phenomena occur.

Table 3.8 Thermal parameters for walls with high internal inertia (κ_1 value) and variable global inertia

High internal inertia solutions ($\kappa_1 > 40$ kJ/m² K)	S1$_{gap+ext}$	S1$_{ext}$	S1$_{gap}$
Description	Expanded clay in the gap and external polystyrene	External insulation layer	Polyurethane foam in existing cavity
	1. Plaster (1.5 cm)	1. Plaster (1.5 cm)	1. Plaster (1.5 cm)
	2. Perf. bricks (8 cm)	2. Perforated bricks (8 cm)	2. Perforated bricks (8 cm)
	3. Cem. mortar (1 cm)	3. Cem. mortar (1 cm)	3. Cem. mortar (1 cm)
	4. Expanded clay (9 cm)	4. Air gap (9 cm)	4. Polyurethane foam (9 cm)
	5. Solid bricks (12 cm)	5. Solid bricks (12 cm)	5. Solid bricks (12 cm)
	6. Adhesive	6. Adhesive	
	7. EPS (9 cm)	7. EPS (9 cm)	
	8. Plaster (0.5 cm)	8. Plaster (0.5 cm)	
Thickness (cm)	41.5	41.5	31.5
U (W/(m² K))	0.29	0.29	0.29
Y_{12} (W/(m² K))	0.009	0.021	0.063
κ_1 (kJ/(m² K))	44 High	40 High	45 High
f [−]	0.032 Low	0.07 Medium	0.217 High

In winter, the curves all present a flat trend for the high value of internal inertia (Fig. 3.15 top graph). The external insulation layer ($S1_{ext}$) and the intervention that combines an interposed insulation with an external layer ($S1_{gap+ext}$) are preferable for the major decrement of the outgoing (in the winter) and entering (in the summer) thermal wave.

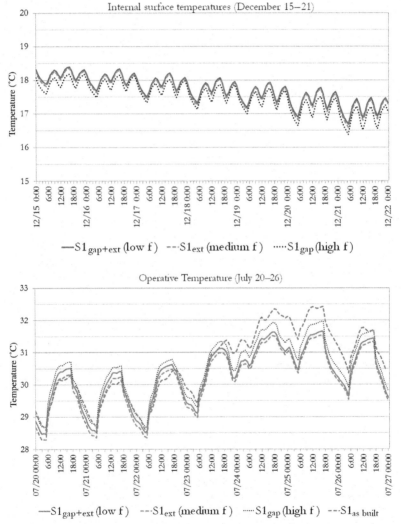

Figure 3.15 High inertia and variable *f*: winter and summer behavior of walls $S1_{gap+ext}$, $S1_{gap}$, $S1_{ext}$ (simulations on the model calibrated on experimental measures).

Table 3.9 Summer and winter discomfort hours for all the walls analyzed

Wall type	κ_1	f	Summer overheating (% discomfort hours)	Winter overcooling (% discomfort hours)
As built	50	0.41	785 (27%)	2678 (92%)
Retrofit horiz. slabs	50	0.41	454 (16%)	2000 (68%)
$S1_{ext}$	40	0.07	420 (14%)	1238 (42%)
$S1_{gap+ext}$	44	0.03	421 (14%)	1178 (40%)
$S1_{gap}$	45	0.2	504 (17%)	2386 (82%)
$S1_{int+gap}$	11	0.08	559 (19%)	1202 (41%)
$S1_{int\ (wood)}$	19	0.06	545 (19%)	1192 (41%)

This simulation regards the retrofit of the horizontal slabs without intervening on the vertical envelope

In summer, the latter intervention presents higher operative temperature (Fig. 3.15 bottom graph) than the former solution for its too low value of decrement factor, but with negligible consequences.

3.5.4 Optimal retrofit intervention from an overall comfort evaluation

The comparison between different traditional energy saving interventions (Table 3.9) confirms that the solutions that optimize the thermal comfort conditions throughout the year are the introduction of an external expanded polystyrene coating ($S1_{ext}$), and the introduction of expanded clay in the gap combined with an external polystyrene layer ($S1_{gap+ext}$). The low value of decrement factor in the latter case determines a summer overheating respect to the traditional coating but the difference is very low (occurring mainly in very hot days) and can be neglected. Hence in the case of an existing insulation in the gap, the adoption of an adjunctive external insulation will have very positive effects on the indoor comfort conditions.

3.6 RETROFIT OF ENVELOPES WITH EXISTING INSULATIONS, HR TYPES

3.6.1 Problem description

To achieve the new requested levels of thermal resistance on existing low insulated building it is important to investigate the state of conservation and the actual performance of the existing insulating materials after a service life of 20−30 years. Besides, the identification of typical degradation phenomena for different insulation materials and for various positions within the envelope could be useful to guide the choice of suitable insulation materials for new constructions.

Consider the real existing solutions illustrated in Table 3.10. Both walls HR1 and HR2 present the insulation on the external side but they are characterized by different type of fabrics behind the insulations, multilayered cavity wall in the former and single layer thermal clay blocks in the latter. HR3 has the insulation layer interposed in the gap of a cavity wall, so it is separated from both the external and internal environment. The wall HR4 has the insulation on the internal side of the mass, realized with dense heavyweight concrete panels with high moisture resistance. All these insulations would have different conservation state depending on the position (more or less influenced by external or internal environment) and on materials adopted.

Very rare are the studies aimed at evaluating the actual envelope conditions in order to select optimal retrofit interventions for the specific case. In recent years, several studies were carried out on the performance decay of hygrometric fibrous insulation materials due to aging. Some researchers have reproduced the actual conditions of full-scale insulated walls portions [20] or single panels of insulating material [21,22] through accelerated aging in climatic chamber. Such studies allow to reach a high accuracy (typical of laboratory measurements), but do not consider the real occupancy and usage conditions. There are very few studies evaluating the long-term performance of the housing buildings taking into account the aspects related to the actual use. Interesting studies focused on the evaluation of one or two aspects, such as the impact of the insulating layer aging on the thermohygrometric behavior on winter and summer consumption [23]. Very rare are the multidisciplinary studies that simultaneously assess the various aspects also including the comfort issue. Some study of our research group regarded single insulation types [18,24].

The present section deepens this aspect and reports the quantification of the actual performance of the insulations in the identified cases. For the worst case (HR4 characterized by the highest f value and the lowest κ_1 value), alternative retrofit solutions are compared through a global convenience evaluation.

Table 3.10 Thermal parameters for four envelopes with existing insulation layers

Envelope with existing insulation	HR1(ext)	HR2(ext)	HR3(gap)	HR4(int)
Description	Unfilled brick cavity wall with external insulation	Clay blocks with ext. spruced plaster (with cork)	Brick cavity wall with existing interposed insulation	Concrete panels with existing internal insulation
Built in	1970	2000	1988	1981
Insulation type	External, EPS panels 4 cm[a]	External, plaster with cork 3 cm	Interposed glass wool 6 cm	Internal glass wool 6 cm
Wall thick. (cm)	40	35	45	23
$U_{th.}$ (W/(m^{2}K))	0.54	0.57	0.45	0.65
U_{exp} (W/(m^{2}K))	0.42	0.96	0.48	0.8
ΔU (W/(m^{2}K))	-22%	$+68\%$	$+6\%$	$+23\%$
Y_{ie} (W/(m^{2}K))	0.05	0.07	0.17	0.34
κ_1 (kJ/(m^{2}K))	53.7	45	50	25
f	0.12	0.12	0.37	0.5

[a]Insulation applied in 1980.

3.6.2 Actual performance of the existing insulation layer depending on material and position

The wall conductance increase/reduction after 20–30 years of service life is defined in the present section by comparing the actual experimental data of on-site thermal transmittance with monitoring values recorded on the same case study several years ago (for the same cases) or with the values certified at the building construction by the manufacturers. Moreover, cylindrical samples of the insulation materials extracted from the walls were compared with new insulations with similar characteristics taken as a reference of the initial conditions.

Table 3.11 compares the values of thermal conductance ($C_{initial}$) at the time of building construction (obtained by experimentations carried out years ago or derived by certifications) and the values measured in-situ during the actual monitorings (C_{actual}). Moreover the table reports the thermal conductivity (λ) obtained from extracted samples of insulation material (λ_{exp}), also compared with the values measured in laboratory for new samples (λ_{decl}). The latter data were also corrected according to the UNI EN ISO 10351 to consider the in-service predicted value (λ_{corr}). The methods are explained in detail in Section 6.4.1.

The retrofitting carried out 30 years ago with the introduction of an external expanded polystyrene (HR1) is still effective from the point of view of thermohygrometric performance since it still guarantees the heat transmission rate values and the elimination of thermal bridges. The heat transmission rate after 30 years seems to have decreased, probably for the progressive reduction in water content in the materials which make up the envelope. In fact this type of insulation thank to its high resistance to

Table 3.11 Performance of insulation materials: at the time of realization and after 20–30 years of service life

	$C_{initial}$ ($\lambda_{decl}/\lambda_{corr}$)	C_{actual} (λ_{exp})	$\Delta C_{initial\text{-}actual}$ ($\Delta \lambda_{corr\text{-}exp}$)
External insulation layer			
HR1	0.5 (−)	0.42 (−)	−16% (−)
HR2	0.6 (0.083/0.085)	1.1 north (0.13 north, 0.15 south)	+46% (+50%)
Interposed insulation			
HR3	0.48 (0.043/0.044)	0.52 (0.047)	+8% (10%)
Internal insulation			
HR4	0.81	0.72	+11%

water at first was used mainly in flotation devices for boats and buoys. Moreover the insulation panels are still effectively bonded to the supporting masonry thank to the high mechanical performance of the anchoring system. On the other hand this solution should be improved from the point of view of the durability of external finishing materials. There is no evidence of chemical deterioration although the surface finishing shows the presence of cracks, corresponding to the joints between the insulation panels (Fig. 3.16). They are due to overheating and differential temperature dilatation; the extent of the damage is greater on the southern side where the temperature swings are more elevate.

The natural plaster with cork (HR2) shows instead a decrement of the insulating performance of about 50%. The performance reduction, more noticeable in the facades with direct exposure to the sun, is underestimated by the $\lambda_{corrected}$ value. Moreover the increasing of thermal conductance of the whole envelope is consistent with this reduction. The material was applied by sprucing and differently by the previous insulation it has maintained not only its adherence but also its initial aesthetical appearance (Fig. 3.17).

As expected the interposed insulation (HR3) demonstrated a more effective conservation of the initial performance for its internal position, but presented the problem of thermal bridges that are easily visible from the external side of the flats and are highlighted in Fig. 3.18 through the aid of a thermocamera. As highlighted in a previous study [24] the apparent density remained constant in time. Thus the glass wool did not change significantly its compaction degree. Moreover the materials did not change water vapor transmission properties with time. Hence the inner porosity distribution, related to the compaction degree, did not change during the service life. Instead the hygroscopic sorption was

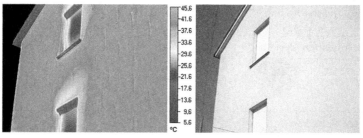

Figure 3.16 Infrared image on a building with 30 years old external polystyrene layer. Southern exposure. *From F. Stazi, C. Di Perna, P. Munafò, Durability of 20-year-old external insulation and assessment of various types of retrofitting to meet new energy regulations, Energy Build. 41 (7) (July 2009) 721–731, ISSN 0378-7788.*

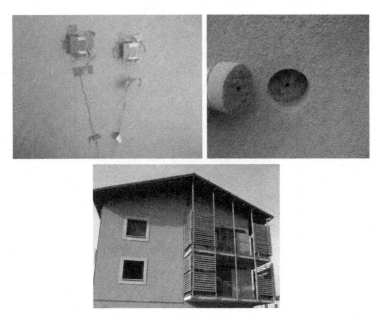

Figure 3.17 Surface conservation state of a building with 30-year-old natural spruced plaster with cork.

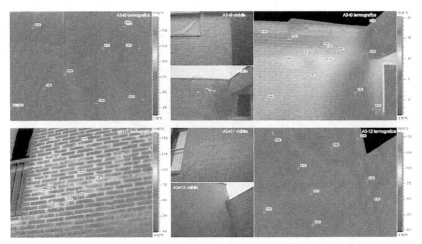

Figure 3.18 Infrared image on a building with 10-year-old interposed rock wool insulation.

found to have increased with aging due to the partial depolymerization for hydrolysis of the organic resin for the presence of moisture. Comparing the measured and declared values an average decrease of thermal conductance of 8% is observed. Moreover the measured values (λ_{exp}) confirm the performance decrease observed in the in–situ thermal transmittance. The obtained results are consistent with the *m* parameter introduced by UNI 10351, representing the thermal conductivity decrease in the average use conditions of the insulation materials which for the glass wool is considered to be about 10%. This reduction of performance could be related to the greater moisture content in the extracted samples (than the new material) due to the water absorption of the material caused by the partial degradation of the organic binder.

The exposure of glass wool insulations at the high humidity of an internal environment (HR4) has almost the same effect highlighted for interposed glass wool insulations (HR3). The results of the thermal transmittance monitoring (Table 3.11) show that the external walls have conductance values higher than the theoretical initial values of 11%.

The thermographic surveys performed on the external envelope show the presence of a thermal bridge at the joints between the prefabricated panels, corresponding to the internal floors and partitions (Fig. 3.19A and B). A considerable heat dispersion was highlighted in the envelope portions below the windows, in correspondence of which are located the internal radiators. A further weakness appears to be the recessed balconies with

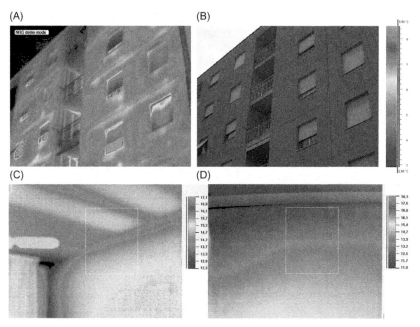

Figure 3.19 Infrared image on a building with 30-year-old internal rock wool insulation and watertight precast concrete panels. (A) Infrared image of the facade, (B) external view of the facade, (C) infrared image of the horizontal roof slab, and (D) infrared image of the first floor slab.

considerable thermal loss. The infrared images from the inside show heat loss both at the ultimate slab of the building (Fig. 3.19C) and on the first floor slab laying above the outside porch (Fig. 3.19D).

3.6.3 Optimal retrofit intervention from a global cost evaluation

The case with existing internal insulation HR4 was selected to identify the best retrofit solution through a global convenience evaluation. In fact, this type of solution presents the worst thermal dynamic parameters having the lowest κ_1 value and the highest f value respect to the other existing HR walls shown in Table 3.10. The experimental data on the selected building (building B.5 in Appendix A) are reported in Section 2.4.

The improvement of the thermal performance was evaluated by introducing three energy efficiency levels (low, medium, and high) for all envelope components (floors, windows, and walls). The analysis included the study of the effects of single interventions or different combinations of them: double interventions (walls and windows or walls and horizontal slabs) and combined interventions (walls, windows, and slabs). Different operating profiles for the heating system were considered (see Section 2.4.4).

Furthermore, only for the walls, alternative interventions were evaluated. The external insulating layer was considered either directly adjacent to the outer side of the existing wall (traditional external insulation layer) or applied in its outer side by leaving an air cavity that could be alternatively closed (in the cold period) or vented (in the hot period through the vents opening), thus configuring a ventilated insulation. Both cases were evaluated by assuming the existing internal insulating layer either to be removed or to be conserved; in the latter case the possibility of changing the internal plasterboard with a high-density clay panel, was also considered. The retrofit measures analyzed are listed in Fig. 3.20.

Table 3.12 reports the thermal characteristics of each studied envelope.

The adoption of high levels of insulation has sometimes a negative effect for overheating. The adoption of thermal barrier solutions with very low f values (0.04), e.g., in W3_h, W4_h, W5_h, W6_h, determines not only higher investment costs than a less insulated solution but also higher energy consumptions for the adjunctive cooling demand that prevails on the saved heating energy.

The graph in Fig. 3.21 and Table 3.13 show that combinations of two or more interventions relative to the envelope (series Double and

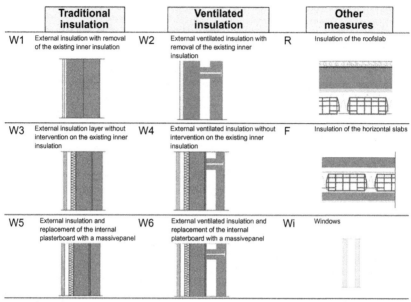

W1: traditional external insulation with removal of the existing inner insulation;
W2: external ventilated insulation with removal of the existing inner insulation;
W3: external insulation layer without intervention on the existing inner insulation;
W4: external ventilated insulation without intervention on the existing inner insulation;
W5: external insulation and replacement of the internal plasterboard with a massive panel;
W6: external ventilated insulation and replacement of the internal plasterboard with a massive panel

Figure 3.20 Sections of the various retrofit measures analyzed.

Combined) are not advantageous solutions since, despite a significant reduction in fuel consumption have a high overall cost.

The optimal interventions regards only the walls (W), by installing exterior dynamic insulation system with either the removal (W2_l, W2_m, W2_h) or the maintenance (W4_m, W4_h) of the inner existing insulation. The minimum point of the curve is represented by the intervention with the outer ventilated insulation and maintaining the existing internal insulation with a medium transmittance level (W4_m) that, despite the slightly lower energy performance than the other solutions provides the lowest global cost (288 €/m^2).

The graph in Fig. 3.22 shows, for the most significant solutions, the overall cost in the case in which there are both the heating and the cooling system, and in the case in which there is only the heating system (excluding the upper quote of the histograms). In the latter case with a

Table 3.12 Main thermal characteristics of different types of walls considered for each level of thermal transmittance

		Insulation thickness t (m)	Thermal transmittance (W/(m² K))	Periodic thermal transmittance (W/(m² K))	Internal areal heat capacity κ_1 (kJ/(m² K))	Decrement factor f
Low level	W1_l	0.08	Wall = 0.36	0.078	45	0.21
	W2_l	0.07	Roof = 0.32	0.085	45	0.21[a]
	W3_l	0.05	Floor = 0.36	0.021	22	0.06
	W5_l	0.05	Window = 2.4	0.022	30	0.06
Medium level	W1_m	0.11		0.057	45	0.21
	W2_m	0.09	Wall = 0.29	0.067	45	0.21
	W3_m	0.07	Roof = 0.26	0.014	22	0.05
	W4_m	0.06	Floor = 0.34	0.017	22	0.05[a]
	W5_m	0.07	Window = 2.0	0.014	30	0.05
	W6_m	0.06		0.017	30	0.05[a]
High level	W1_h	0.16		0.039	45	0.19
	W2_h	0.14	Wall = 0.20	0.044	45	0.19[a]
	W3_h	0.12	Roof = 0.20	0.008	22	0.04
	W4_h	0.11	Floor = 0.20	0.009	22	0.04[a]
	W5_h	0.12	Window = 1.3	0.008	30	0.04
	W6_h	0.11		0.01	30	0.04[a]

[a] f value obtained by considering the air gap closed.

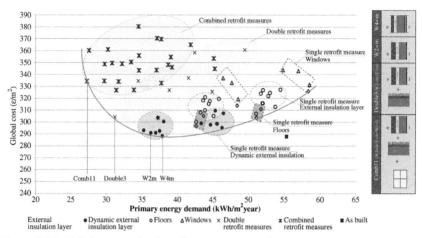

Figure 3.21 Global cost evaluation of alternative retrofit interventions.

Table 3.13 Winter and summer consumptions (continuous operation), energy performance, and global cost for selected interventions

Single intervention	Winter consumption (kWh/m² year)	Summer consumption (kWh/m² year)	Energy performance index (kWh/m² year)	Global cost (€/m²)
W2_l	22.99	6.42	36.9	291
W2_m	22.43	6.36	36.2	291
W2_h	21.62	6.27	35.2	293
W2_l + Floor_l + Window_l	15.95	7.27	31.7	331
W3_l	23.08	10.14	45.1	305
W3_m	22.44	9.31	42.6	300
W3_h	21.66	10.24	43.9	301
W3_l + Floor_l + Window_l	16.04	10.68	37.8	338
W4_m	22.47	7.09	37.9	288
W4_h	21.66	7.29	37.5	292
W4_m + Floor_m + Window_m	12.84	8.93	32.2	345
W4_h + Floor_h + Window_h	8.66	9.94	30.2	358

natural ventilation in summer phase, the various retrofit solutions led to variable comfort levels that could be compared as the percentage of discomfort hours over the entire season (*dashed line*, referred to secondary *y*-axis).

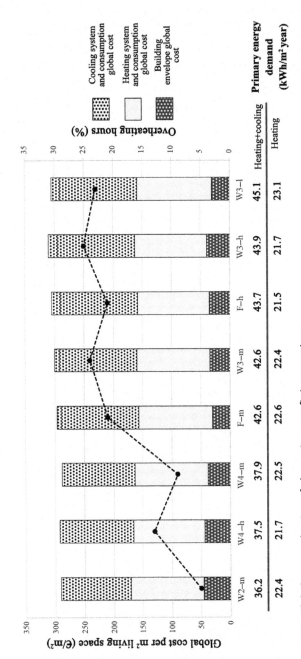

Figure 3.22 Global cost evaluation of alternative retrofit interventions.

Insulation of the floors (F_m, F_h) and insulation of the vertical envelope with the traditional external insulation layer (W3_l, W3_m, W3_h) are, from an overall assessment, more disadvantageous for the highest consumptions in summer phase or, in the case of a passive use of the accommodation, for the highest levels of discomfort.

Between the other three solutions, W4_m (ventilated external insulation layer without removing the existing internal insulation) is preferable because it guarantees the lowest overall cost mainly due to an initial investment lower than W2_m (where there is the removal of the insulation inside) and then W4_h (that has a greater thickness of insulation). Moreover, the solution W4_m presents high levels of comfort if adopting natural ventilation during summer (i.e., excluding the upper histogram portion), thanks to its low (but not excessively low) f value.

The graph in Fig. 3.23 explains how the best solution (W4_m, the ventilated insulation) is able to guarantee appropriate comfort levels. It reports the study of the internal surface temperatures during the hottest summer period (July 18−28) of W4_m compared with the initial as built condition and with a traditional intervention of external insulation (W3_m). Differently by other existing walls (e.g., solid masonry in Section 3.4.2) all the retrofit interventions on such precast concrete panel

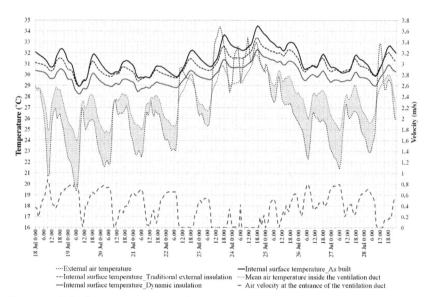

Figure 3.23 Surface temperatures of intervention W3_m and W4_m compared with as built condition.

reduce the internal temperatures respect to the initial "as built" situation without overheating drawbacks for the moderate (not excessively low) f value of this initial envelope solution. The ventilated system values stand lower than the other intervention curves. The thermal behavior of such system strictly depends from the temperature difference between the outside air and the air within the channel, which is the main driving force for the stack effect activation inside the cavity. The study of the air velocity values inside the channel highlights that the ventilation is effectively activated when the channel air temperature considerably exceeds the outside air temperature value. This happens (*shaped area between dotted lines*)

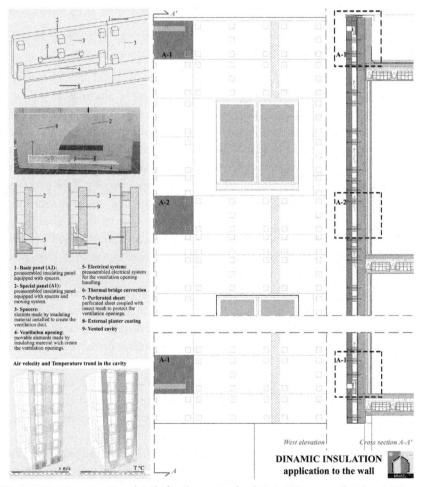

Figure 3.24 Construction details for the optimal solution W4_m, see Section 3.6.3.

during the whole day and particularly in nighttime for most of the represented period with typical summer temperature conditions, while the ventilation is not effective on extremely hot days (from July 22 to 25). The benefits of adopting a vented system respect to a traditional external insulation could be quantified in a reduction of 1°C in typical summer days and a slightly lower performance on the days with extremely high temperatures.

An example of the constructive details for the optimal vented solution W4_m is shown in Fig. 3.24.

3.7 DESIGN PATTERNS

During the summer months, a building with solid brick masonry has high comfort levels thanks to the high capacity of the masonry (low f value) which mitigates the outdoor climate conditions. However, this advantageous feature in the summer as built condition creates an overheating problem after a retrofit with high insulation thicknesses. This is due to the too low f value reached after the insulation measure (down to 0.003), that turn the external envelope in a thermal barrier. This problem is instead negligible in lightweight existing envelopes, such as semisolid masonries or cavity walls, for the higher initial f value of those configurations. For the very massive solutions, ventilated insulation could prevent the summer overheating, reducing the temperatures down to 3°C respect to traditional external insulation, while for lightweight load baring walls and for stratified ones the reduction is less noticeable, being about 1°C.

Regarding the performance of the insulation materials after a long service life, the external spruced cork plaster is not durable especially in climates with aggressive radiation, while the expanded polystyrene even if placed on the external side is highly durable but some measure should be studied to avoid the external finishing cracking (e.g., through nanoparticle additives or external sun-screens). The glass wool insulation either interposed or paneled on the internal side has a performance decrease of about 10%.

From a global cost evaluation, the existing insulations should not be removed but combined with the introduction of an adjunctive insulation layer on the external side, possibly leaving an openable vented cavity to enhance the summer performance. The other building components could be advantageously maintained in the as built condition, i.e., without the introduction of insulation layers.

REFERENCES

[1] E. Reid, Understanding Buildings: A Multidisciplinary Approach, MIT Press, Cambridge, 1984, ISBN: 0-262-68054-8.
[2] B. Givoni, Climate Consideration in Building and Urban Design, John Wiley and Sons, New York, USA, 1998, ISBN 0-471-29177-3.
[3] M. Grosso, Design guidelines and technical solutions for natural ventilation, Natural Ventilation in Buildings—A Design Handbook, F. Allard; James & James (GBR), London, 1998, pp. 195—254, ISBN: 1873936729.
[4] A.S. Dili, M.A. Naseer, T. Zacharia Varghese, Passive control methods of Kerala traditional architecture for a comfortable indoor environment: a comparative investigation during winter and summer, Build. Environ. 45 (2010) 1134—1143.
[5] Y. Ryu, S. Kim, D. Lee, The influence of wind flows on thermal comfort in the Daechung of a traditional Korean house, Build. Environ. 44 (2009) 18—26.
[6] F. Wang, Y. Liu, Thermal environment of the courtyard style cave dwelling in winter, Energy Build. 34 (2002) 985—1001.
[7] R. Ooka, Field study on sustainable indoor climate design of a Japanese traditional folk house in cold climate area, Build. Environ. 37 (2002) 319—329.
[8] N. Cardinale, G. Rospi, A. Stazi, Energy and microclimatic performance of restored hypogeous buildings in south Italy: the "Sassi" district of Matera, Build. Environ. 45 (2010) 94—106.
[9] Z. Yilmaz, Evaluation of energy efficient design strategies for different climatic zones: comparison of thermal performance of buildings in temperate-humid and hot-dry climate, Energy Build. 39 (2007) 306—316.
[10] J.A. Orosa, T. Capente, Thermal inertia effect in old buildings, Eur. J. Sci. Res. 27 (no.2) (2009) 228—233.
[11] F. Stazi, F. Angeletti, C. Di Perna, Traditional houses with stone walls in temperate climates: the impact of various insulation strategies, in: A. Almusaed (Ed.), Effective Thermal Insulation—The Operative Factor of a Passive Building Model, InTech, 2012, pp. 45—60.
[12] S. Martín, F.R. Mazarrón, I. Cañas, Study of thermal environment inside rural houses of Navapalos (Spain): the advantages of reuse buildings of high thermal inertia, Construction Build. Mater. 24 (2010) 666—676.
[13] K.J. Kontoleon, D.K. Bikas, The effect of south wall's outdoor absorption coefficient on time lag, decrement factor and temperature variations, Energy Build. 39 (2007) 1011—1018.
[14] H. Asan, Investigation of wall's optimum insulation position from maximum time lag and minimum decrement factor point of view, Energy Build. 32 (2000) 197—203.
[15] E. Kossecka, J. Kosny, Influence of insulation configuration on heating and cooling loads in a continuously used building, Energy Build. 34 (2002) 321—331.
[16] P.T. Tsilingiris, Wall heat loss from intermittently conditioned spaces—the dynamic influence of structural and operational parameters, Energy Build. 38 (2006) 1022—1031.
[17] C. Di Perna, F. Stazi, A. Ursini Casalena, M. D'Orazio, Influence of the internal inertia of the building envelope on summertime comfort in buildings with high internal heat loads, Energy Build. 43 (2011) 200—206.

[18] F. Stazi, C. Di Perna, P. Munafò, Durability of 20-year-old external insulation and assessment of various types of retrofitting to meet new energy regulations, Energy Build. 41 (2009) 721–731.
[19] M. Hall, Materials for Energy Efficiency and Thermal Comfort in Buildings, Woodhead Publishing Series in Energy, 2010.
[20] Z. Pavlík, R. Černý, Experimental assessment of hygrothermal performance of an interior thermal insulation system using a laboratory technique simulating on-site conditions, Energy Build. 40 (2008) 673–678.
[21] R. Peuhkuri, C. Rode, K.K. Hansen, Non-isothermal moisture transport through insulation materials, Build. Environ. 43 (2008) 811–822.
[22] F. Björk, T. Enochsson, Properties of thermal insulation materials during extreme environment changes, Construction Build. Mater. 23 (2009) 2189–2195.
[23] L.F. Cabeza, A. Castell, M. Medrano, I. Martorell, G. Pérez, I. Fernández, Experimental study on the performance of insulation materials in Mediterranean construction, Energy Build. 42 (2010) 630–636.
[24] F. Stazi, F. Tittarelli, G. Politi, C. Di Perna, P. Munafò, Assessment of the actual hygrothermal performance of glass mineral wool insulation applied 25 years ago in masonry cavity walls, Energy Build. 68 (2014) 292–304.

CHAPTER FOUR

New Envelopes

4.1 INTRODUCTION

The 2012/27/UE Directive has set a target for all new buildings to be nearly zero-energy by 2020. The reaching of this goal makes it necessary to adopt not only renewable sources but also highly efficient envelopes. The text highlights that thick insulation layers combined with high performing new materials and the new tendency to multiply the layers have led to the introduction of a new type of envelope in which the decrementing attitude is strongly elevated, thus configuring thermal barriers. Such envelopes could have overheating drawbacks.

This chapter introduces the main new envelope typologies subdivided in two families, lightweight (platform-framed envelopes) and massive load bearing (cross-laminated timber (CLT) envelopes, wood-cement solutions, clay masonries), and explores their different behavior through experimental measures and analytical extensions. Optimization solutions are identified in various climates to maximize energy saving, comfort and global costs.

4.2 LIGHTWEIGHT (L) AND MASSIVE (M): TYPES OF RECENTLY ADOPTED ENVELOPES

The new envelopes recently diffused in the European countries could be grouped in two typologies (Table 4.1):

- *Lightweight framed envelopes*: among the common methods of framing a house, platform framing has become predominant.
- *Massive load-bearing walls*: very different walls fall within this category, such as the load-bearing CLT structures with insulating layers on both side of the timber, the wood-cement walls with interposed insulations, the masonries with clay units and external insulating layers.

Thermal Inertia in Energy Efficient Building Envelopes.
DOI: http://dx.doi.org/10.1016/B978-0-12-813970-7.00004-2

© 2017 Elsevier Inc.
All rights reserved.

127

Table 4.1 New envelope types commonly adopted in Europe

Lightweight (L series)		Massive load bearing (M series)	
Platform frame		Cross-lam. timber−Wood-cement−Masonry	
Thickness = 43 cm		Thickness = 28−43−49 cm	
$U = 0.12$ W/(m^2 K)		$U = 0.26-0.26-0.26$ W/(m^2 K)	
$\kappa_1 = 24$ kJ/(m^2 K)		$\kappa_1 = 20-37-43$ kJ/(m^2 K)	
$f = 0.09$		$f = 0.16-0.027-0.038$	

The right side is the external side.

Each of these solutions was generally born as an evolution of one traditional envelope, modified by introducing adjunctive layers and adopting new thermal performing materials. However, these modifications often occurred without the parallel development of a technical culture to adapt the solutions to the specific climates and use.

The *lightweight framed* constructions, e.g., were conceived in cold climates and then adopted even in warm Mediterranean countries. Here they could suffer problems of thermal control as highlighted in the previous chapters, since they cannot take advantage from the thermal inertia. Among this construction type, the "platform frame" is the most common method. The name derives from the way of construction according to which the floor structures bear onto load-bearing wall panels, thereby creating a "platform" for construction of the next level of wall panels. Timber wall panels are comprised of vertical studs at regular spacing (typically 600 mm) and sheathing wall panels (commonly plywood panels or composite wood skins) that act as bracings thus providing resistance to horizontal actions. Insulation of the building envelope typically occurs within the stud wall spaces. Recently this kind of construction evolved to satisfy new energy saving requirements by strongly increasing the insulation thickness (using wider studs to provide space for more insulation) and introducing other insulating layers on both sides of the timber paneled structure. Other adjunctive layers were also added in the last decades using new performing and highly specialized materials, such as a vapor retarder on the warm side of the insulation and a vapor-permeable membrane on the outer face of the panel. The interior wall finish is typically a gypsum board.

Within the massive walls, a new product category of wooden envelopes known as *massive timber* is included. It is made from several layers of lumber board (generally an odd number), stacked crosswise (usually at 90 degree angles) and glued together on their wide faces. This cross lamination provides dimensional stability, strength, and rigidity, making CLT a viable alternative to masonry. CLT has been popular in Europe for more than 20 years, with extensive research supporting its widespread use. The new energy saving regulations have determined the introduction of a thick insulation layer on the external side of the load-bearing timber wall, while the insulation in the internal side is usually introduced to host the plants pipes.

Among the new massive techniques, the *wood-cement solutions* are increasingly used in European countries. In North America, it has been in production since the early 1950s. The wood is chipped into wood fiber, mineralized and bonded together with Portland cement. The obtained material is then molded into shapes and the insulation material is embedded in each block. The blocks are subsequently filled with concrete, so the internal filling presents mediated properties among those of wood-cement, concrete, and expanded polystyrene (EPS) insulation. These new solutions have very high attenuating attitude, and are therefore characterized by a strongly decoupled behavior between the external and the internal side. They behave as thermal barriers, thus blocking not only the incoming but also the outgoing heat flux. This behavior is beneficial in "stationary" cold climates when the main scope is to hinder the heat loss, but it could create overheating in the case of high internal gains.

The well-known *masonry* techniques have recently adopted very thick insulation layers combined with new highly efficient materials, such as new specially designed light hollow clay blocks and foamed polymeric insulations to reach the requested very high thermal resistances. This has introduced a new kind of hyperinsulated building wall, with possible overheating phenomena.

Nevertheless, a proper mix of light and heavy components, a careful use of the sun shadings, an effective ventilation, a proper insulation of the external envelope, an accurate linings choice and, perhaps of greater importance, a deep feeling of limitation and consequences of each technique, will produce comfortable and sustainable buildings for almost all the new introduced envelope solutions.

To contribute to the debate regarding these open questions the present chapter reports the results of experimental investigations and analytical simulations for different types of new envelopes, representative of the identified typologies. Optimized solutions are drawn for each envelope through comfort, energy, and environmental analysis also considering the behavior at the varying of some parameters: climates, profiles for heating/mechanical ventilation plants, shadings. The variations regarded not only the vertical walls but also, simultaneously, the horizontal external floors (roof and ground floor slabs) by adopting the same technology selected for the vertical walls.

4.3 LIGHT-FRAMED CONSTRUCTIONS

4.3.1 Problem description

Experimental measures on superinsulated platform-framed envelope are very rare in hot-summer climates. In international literature the most common studies on light-framed high energy efficient buildings are generally analytical, aimed at the definition of project specifications [1,2] and evaluation methodologies [3,4], or focused on the determination of the energy performance of single cases [5,6]. Moreover, since originally conceived for cold climates, they are commonly focused on the optimization of the winter energy saving and only rarely on the summer overheating [7,8]. Furthermore, the abovementioned studies generally omit life cycle assessment (LCA) and few researchers stress the contradiction between the better energy performance achieved through thick insulations and the higher environmental impacts due to the increase in whole-life energy costs of building construction [9–11].

The very few experimental studies that provide real data on passive houses are carried out in cold North-American and North-European climates [12,13] or in extreme arid ones [14,15]. Moreover, the extension to other situations through complex simulations on virtual models calibrated through measures is uncommon.

In the present work, a real case study (building C.3 in Appendix A) has been monitored and simulated. The summer comfort levels and the environmental impacts are quantified and appropriate passive strategies to reduce overheating and optimize the behavior in terms of energy performance, thermal comfort, and sustainability are identified.

Consider the dynamic thermal parameters of the external envelope reported in Table 4.2. The thermal transmittance is extremely low for all the constructive elements. The vertical walls have low values of both κ_1 and decrement factor f. The other envelope components (ground and roof floors) assume medium values.

4.3.2 Comfort, experimental measures

A summer experimentation on the selected building shows the presence of overheating due to the very low thermal transmittance values and highlights that in high thermal resistant and airtight envelopes the majority of the heat dispersion occurs toward the ground.

The operative temperature analysis (Fig. 4.1) of the southeast rooms placed on different building levels highlights the presence of overheating on the first floor (with values constantly around $28-29°C$) and temperatures subjected to greater fluctuations on the ground floor (values between $26.5°C$ and $28.5°C$). This different regime of indoor temperatures also influences the relative humidity values. In fact, the lower is the air temperature, the higher will be the relative humidity due to the solubility decrease of water vapor in the air. The values on the ground floor stand on average on $55\%-60\%$, while on the first floor on $45\%-50\%$, remaining in both cases within the limits set by EN 15251 (lower limit of 25% and upper limit of 60% for new buildings in category II).

At the ground floor level, the fluctuations are due to the great dispersion toward the underlying ground, as demonstrated by the comparison between the heat fluxes analysis through all constructive elements of the external envelope, roof, wall, and floor (Fig. 4.2). The graph shows that the highest fluxes occur on the ground floor slab with very high and always negative values, indicating a constantly outgoing trend. In the roof, instead there is an always-entering flux because of, the inside temperatures are always lower than the surface ones for the temperature stratification in the internal environment. This is clearly visible in the graph in Fig. 4.3, regarding the internal surface temperatures at different wall heights.

An experimental comparison was made by choosing 2 days with similar climatic conditions (August 31 and September 2) and adopting either

Table 4.2 Light-framed envelope of building C.3 (see Appendix A)

External wall	t (cm)	Floor on the ground	t (cm)	Roof	t (cm)
Internal side		*Internal side*		*Internal side*	
Double plasterboard	2.6	Ceramic tiles	1	Fir planking	3.3
Rockwool insulation	5	Base	1.5	Vapor barrier	0.2
OSB panel	1.8	OSB panel	1.8	Rockwool insulation	30
Vapor barrier	0.25	Screed	10	OSB panel	1.25
Rockwool insulation[a]	16	Lightweight screed	4	Transpiring sheet USB	0.2
Vapor barrier	0.25	OSB panel	1.8	Air chamber	5
OSB panel	1.8	XPS insulation	16	OSB panel	1.25
Rockwool insulation	10	Reinforced concrete slab		Bituminous sheath	0.5
Air chamber	4	*External side*		*External side*	
Weather-proof cement	1.3				
External side					
	43.0		36.1		41.7

	External wall	Floor on the ground	Roof
U (W/(m^2 K))	0.12	0.18	0.12
κ_1 (kJ/(m^2 K))	24.3	43.8	35.4
f (−)	0.092	0.106	0.154
Δt (h)	14	12	16
Y_{12} (W/(m^{2}K))	0.011	0.019	0.018

[a]The rockwool insulation (16 cm) is interposed between the struts with a pace of 62.5 cm.

Source: Super-insulated wooden envelopes in Mediterranean climate: summer overheating, thermal comfort optimization, environmental impact on an Italian case study, Energy Build. 138 (March 1, 2017) 716–732, ISSN 0378-7788.

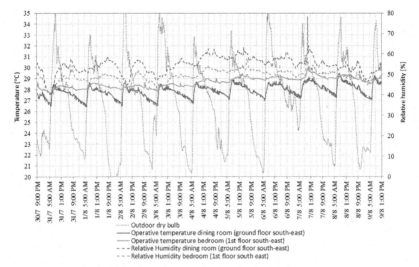

Figure 4.1 Thermal comfort summer experimental measures on two rooms on the southern side at the ground floor and first floor levels. *From Super-insulated wooden envelopes in Mediterranean climate: summer overheating, thermal comfort optimization, environmental impact on an Italian case study, Energy Build. 138 (March 1, 2017) 716–732, ISSN 0378-7788.*

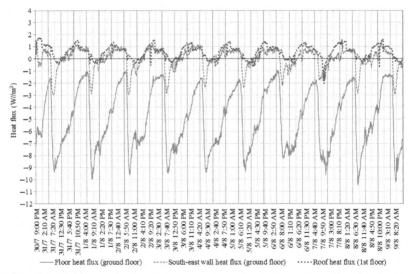

Figure 4.2 Summer experimental measures on heat fluxes through the envelope and through the ground floor slab. *From Super-insulated wooden envelopes in Mediterranean climate: summer overheating, thermal comfort optimization, environmental impact on an Italian case study, Energy Build. 138 (March 1, 2017) 716–732, ISSN 0378-7788.*

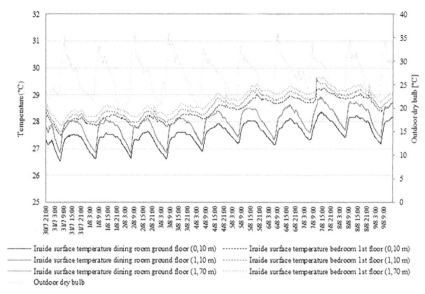

Figure 4.3 Summer experimental measures on internal surface temperatures at different wall heights.

natural ventilation or controlled mechanical ventilation (CMV) with night free-cooling (Fig. 4.4). The building in fact has a CMV system with heat recovery, equipped with a bypass valve for free cooling at night (see the scheme in Fig. 4.4). To evaluate the natural ventilation efficacy, the mechanical ventilation plant was switched off for 1 day and the windows of the ground floor were kept open for most of the day and during the night, as shown by the measured values of inside air velocity. This natural vented condition shows a better behavior than the CMV in the selected hot days. However, simulations under equal external condition are necessary (see Section 4.3.5.2).

4.3.3 Consumptions

The winter and summer conditioning consumptions obtained are very low (Table 4.3).

The summer consumption has values very close to the winter ones.

The study of the thermal bridges (Table 4.4) reveals that in lightweight framed constructions an almost uniform thermal resistance is achieved.

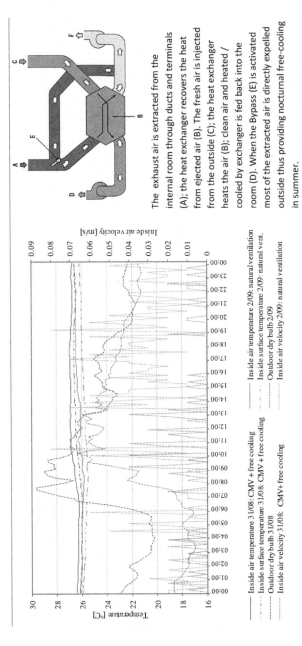

The exhaust air is extracted from the internal room through ducts and terminals (A); the heat exchanger recovers the heat from ejected air (B). The fresh air is injected from the outside (C); the heat exchanger heats the air (B); clean air and heated / cooled by exchanger is fed back into the room (D). When the Bypass (E) is activated most of the extracted air is directly expelled outside thus providing nocturnal free-cooling in summer.

Figure 4.4 Experimental comparison between two similar days adopting different cooling strategies, mixed mode ventilation (mechanical ventilation during the day and free cooling at night) and natural ventilation. Scheme of the controlled mechanical ventilation system with heat recovery and bypass for night free cooling.

Table 4.3 Primary energy consumptions, continuous heating, analytic simulations

Winter (kWh/m² year)	Summer (kWh/m² year)
6.47[a]	4.03[b]

[a]COP = 3.1.
[b]EER = 2.9.

Table 4.4 Thermal bridges, numerical simulations

Type of thermal bridge/linear thermal transmittance	Ψ_i (W/m K)
C1—corner of external walls	0.02
IF1—intermediate floor/external brick wall	0.02
B—balcony	0.01
GF—ground floor/external plastered wall/roof of unheated space	0.05
R5—roof/external wall	0.02

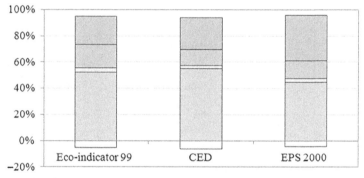

☐ Construction ☐ Transport ☐ Maintenance ☐ Use (summer + winter) ☐ End-of-life

Figure 4.5 Life cycle phase results of the building during 75-year life span with Eco-indicator 99, CED, and EPS 2000 methods.

4.3.4 Environmental sustainability

The study of the impacts distribution in the various life cycle phases (Fig. 4.5) with Eco-indicator 99, CED, and EPS 2000 methods (weighting analysis) shows that the greatest total damage is mainly due to the initial "construction" phase according to all methods.

The "use" phase shows a significant impact due mainly to the summer cooling achieved by burning fossil fuels. Fossil fuels constitute a significant repository of carbon buried deep underground. Burning them results in the conversion of this carbon into carbon dioxide, which is then released into the atmosphere. Instead, the "transport" and "end-of-life" phases affect only a small part compared to the building overall impact. In particular, concerning the "end-of-life," having hypothesized the recycle of some materials (aluminum, glass, PVC, concrete, etc.), the environmental burdens associated with manufacturing of the substituted product are subtracted from the system. The negative value (net absorption of emissions) of the balance between environmental impacts and gains indicates that the avoided impacts are greater than those caused. The wood results to be creditor in terms of CO_2-Eq because the balance takes into account the CO_2 stored by wood during its life cycle.

The graph in Fig. 4.6 reports the evaluation with Eco-indicator 99 focusing on the construction phase, which came out the one with the biggest weight on global LCA and concentrating on materials with the greatest environmental impact. The graph shows that the highest impact could be assigned to the insulation materials that present the greatest harm to "Human Health." In particular, the most impacting materials are rockwool insulations that are present in large quantities in the building.

Regarding basic materials characterizing the envelope and the relative structure, the OSB panel has the greatest environmental burden mainly affecting "Resources" category for the energy used for the complex production process of the material that comprises the hot pressing. The finishes have a very low impact apart from PVC and bituminous sheaths, respectively, impacting on human health and resources.

Table 4.5 shows the life cycle impact assessment (LCIA) during 75-year life span reporting data according to the characterization phase (category indicators) and the overall results of each method in relation to the weighting analysis (Pt). The relative impacts for each category are not comparable since they have different units. However, the detailed study of characterized values allows identifying the main problems for each category before the weighting of the burdens through discretional weighting factors (see the relative discussion on Chapter 6: Experimental Methods, Analytic Explorations, and Model Reliability).

The values with Eco-indicator 99 method show that the building has a high environmental impact in "Resources" category (2.29E + 05 Pt) and in particular in Fossil fuels subcategory. This is due to the summer cooling achieved by burning fossil fuels and for the production processes of the rockwool insulation (the volcanic rock is melted, transformed into fibers and sprayed with resin and oil, then sent to polymerization furnaces), present in large quantities in the technique. Also in "Ecosystem Quality" the building presents high environmental burdens, in particular as regard the wood extraction from forests and green areas (Land use).

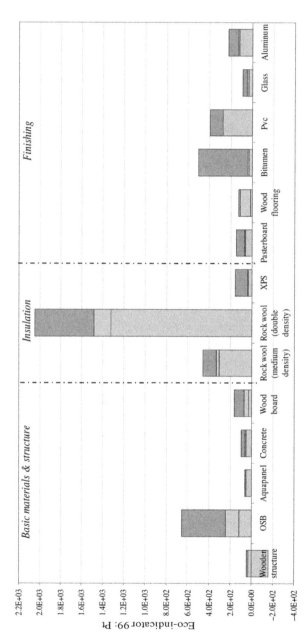

Figure 4.6 Construction phase results of the building at year 0 with Eco-indicator 99 method. *From Super-insulated wooden envelopes in Mediterranean climate: summer overheating, thermal comfort optimization, environmental impact on an Italian case study, Energy Build. 138 (March 1, 2017) 716–732, ISSN 0378-7788.*

Table 4.5 LCIA: results of passive house during 75-year life span with Eco-indicator 99, CED, EPS 2000, and IPCC 2001 GWP methods

LCA method/category	Units	
Eco-indicator 99 (H)—hierarchist perspective		
Human health	**DALY**	**1.47E-01**
Carcinogens	DALY	3.68E-02
Respiratory organics	DALY	1.32E-04
Respiratory inorganics	DALY	1.00E-01
Climate change	DALY	9.82E-03
Radiation	DALY	3.66E-04
Ozone layer	DALY	1.97E-05
Ecosystem quality	**PDF m² year**	**1.28E + 04**
Ecotoxicity	PDF m² year	4.68E + 03
Acidification/eutrophication	PDF m² year	1.90E + 03
Land use	PDF m² year	6.22E + 03
Resources	**MJ surplus**	**2.29E + 05**
Minerals	MJ surplus	3.79E + 03
Fossil fuels	MJ surplus	2.25E + 05
Total	**Pt**	**1.20E + 04**
Cumulative Energy Demand—CED		
Not renewable sources	**MJ-Eq**	**2.06E + 06**
Fossil	MJ-Eq	1.90E + 06
Nuclear	MJ-Eq	1.59E + 05
Renewable sources	**MJ-Eq**	**4.61E + 05**
Biomass	MJ-Eq	4.19E + 05
Wind, solar, geothermal	MJ-Eq	2.48E + 03
Water	MJ-Eq	3.97E + 04
Total	**Pt**	**2.52E + 06**
Environmental Priority Strategy—EPS 2000		
Human health	**Person Year**	**2.61E + 00**
Life Expectancy	Person Year	1.62E-01
Severe Morbidity	Person Year	1.67E-02
Morbidity	Person Year	3.57E-02
Severe Nuisance	Person Year	6.46E-02
Nuisance	Person Year	2.33E + 00
Ecosystem production capacity	**kg**	**2.10E + 03**
Crop growth capacity	kg	2.86E + 02
Wood growth capacity	kg	− 2.27E + 03

(*Continued*)

Table 4.5 (Continued)

LCA method/category	Units	
Fish and meat production	kg	$-1.02E+01$
Soil acidification	H + eq.	$5.65E+02$
Prod. cap. irrigation water	kg	$1.76E+03$
Prod. cap. drinking water	kg	$1.76E+03$
Abiotic stock resources	**ELU**	**6.59E+04**
Depletion of reserves	ELU	$6.59E+04$
Biodiversity	**NEX**	**1.19E-09**
Species Extinction	NEX	1.19E-09
Total	**Pt**	**8.27E+04**
Global potential warming—IPCC 2001 GWP		
IPCC GWP 100a	**kg $_{CO2}$-Eq.**	**4.78E+04**

The study with CED method confirms the high environmental impacts caused by the "Not Renewable Sources," and in particular by Fossil for the reasons given for the previous method.

The same is confirmed also with EPS 2000 method in the "Abiotic Stock Resources" category due to exploitation and exhaustion of reserves. The CO_2 emissions detected with IPCC 2001 GWP method are mainly due to the Fossil fuel combustion.

4.3.5 Optimized solutions

4.3.5.1 Increase of the internal inertia

The present section identifies the most suitable internal linings to enhance comfort levels with low cost, having fixed a very low thermal transmittance value and thus considering very similar energy saving performance.

Consider the walls in Table 4.6 in which different solutions for the inner lining are compared, increasing the internal heat capacity. The solutions regard the introduction on the internal side of: wood fiber panels; plywood panels; dry clay panels, in single and double layers; dry clay bricks (11.5 cm thick); a lime-cement plaster; wood–cement panels; solid bricks 5.5 cm thick; solid bricks 12 cm thick. The walls are ordered with increasing values of internal inertia (κ_1 values).

The comparison of the envelope with the different internal linings on a southeast exposed room on the ground floor building level showed (Fig. 4.7) that the best constructive solution is the insertion of traditional

Table 4.6 Inner lining alternatives

Internal lining	Low capacity			Medium capacity			High capacity	
	Wood fiber panel	Plywood panel	Clay panel(s)	Dry clay brick	Lime–cement plaster	Wood–cement panels	Solid bricks	Thick solid bricks 12 cm
	Wood fiber 10 cm	Insul. 5 cm / Plywood 4cm	Insul. 5 cm / Clay panel 2.2 cm	Insul. 5 cm / Clay brick 11.5 cm	Insul. 5 cm / Plaster 3 cm	Insul. 5 cm / Wood–cement 4 cm	Insul. 5 cm / Solid brick 5.5 cm	Insul. 5 cm / Solid brick 12 cm
Thick. (cm)	45.4	44.4	42.6/44.8[a]	51.9	43.4	44.4	45.9	52.4
U (W/(m²K))	0.12	0.12	0.12/0.12[a]	0.12	0.12	0.12	0.12	0.12
κ_1 (kJ/(m²K))	16.8	26.8	29.1/46.2[a]	42.7	47	58.2	60.8	68.9
f (−)	0.008	0.078	0.093/0.08[a]	0.036	0.087	0.058	0.071	0.037
Y_{12} (W/(m²K))	0.001	0.009	0.012/0.01[a]	0.004	0.011	0.007	0.009	0.005
λ (W/(m K))	0.048	0.13	0.47	0.25	0.9	0.26	0.78	0.78
c (J/(kg K))	2100	1600	1000	720	1000	1880	940	940
ρ (kg/m³)	265	500	1300	1350	1800	1350	1700	1700

[a]The values after the bar (/) are referred to the application of a double layer of clay panels (for a total thickness of 4.4 cm).

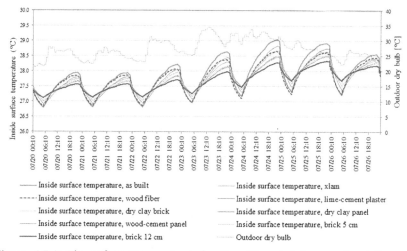

Figure 4.7 Inside surface temperatures for the different linings under daily controlled mechanical ventilation and night free cooling (results from analytical simulations). *From Super-insulated wooden envelopes in Mediterranean climate: summer overheating, thermal comfort optimization, environmental impact on an Italian case study, Energy Build. 138 (March 1, 2017) 716–732, ISSN 0378-7788.*

solid bricks (12 cm) in the internal side of the wall. This solution in fact presents the highest internal thermal inertia (κ_1 value of about 70 kJ/m^2 K) combined with a high decrementing attitude of the incoming thermal wave (*f* value of 0.037). This value is not too low, while it is excessively low for the solution with wood fiber panels that creates overheating phenomena. Inversely in the solid brick solution this value is not too high, as for the lime-cement plaster that has the worst behavior on the reduction of incoming heat gains, especially in the hottest summer days. The selection of the identified optimal dynamic parameters guarantees an improvement respect to the as built wall with a surface temperatures lowering of 1.5°C in the best case and fluctuations attenuation, ensuring better comfort conditions. The adoption of a massive interior finishing is most important in extremely hot days (July 23–26). However, it is important to highlight that no solution reaches comfortable temperatures making necessary to combine the adoption of massive linings with other cooling strategies such as ventilation (see Section 4.3.5.2).

Figs. 4.8 and 4.9 report an experimental evidence on the benefices given by adopting an inner massive lining in a lightweight framed solution.

The experimentation was carried out in a platform-framed building in which the owner (based on the results of the present research) introduced a 5 cm thick solid brick layer on the inner side of the wall

Figure 4.8 Internal massive linings mounted on wooden envelopes based on the results of the present research.

(Fig. 4.8). A small portion of the walls was left without the massive lining and a double plasterboard layer was mounted to have a reference case. The two walls were on the same room, to compare their behavior under the same both internal and external environment. The inner room was not occupied during the survey.

The internal surface of the solid brick lining recorded just at the end of the cold season, resulted to be constantly 1°C warmer than the surface of the plasterboard since the former has stored the heat (Fig. 4.9). This demonstrates that the massive material enables the envelope to behave as a high capacity solution.

Another experimental evidence of the benefice of adopting an inner massive lining on the platform-framed technique is reported in Appendix A, case C.1.

4.3.5.2 Ventilation strategies

The numerical simulations made it possible to explore the benefices of alternative ventilation types, also combined with the different inner linings. The following cooling strategies are compared:

- CMV + free cooling: it combines CMV set at 0.5 air changes per hour all the time and free cooling activated during the night from 11:00 p.m. to 7:00 a.m. assuming a temperature set-point of 26°C. This case represents the as built condition.

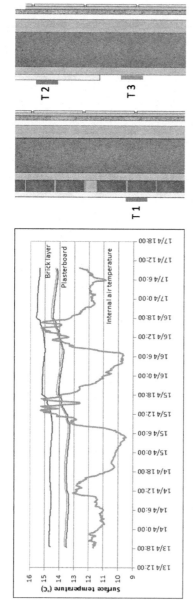

Figure 4.9 Experimental measures on a lightweight framed envelope with and without inner massive lining.

- Hybrid ventilation: it is conceived as CMV during the day (6:00 a.m.–6:00 p.m.), fixed at 0.5 ach, and natural ventilation during the night (6:00 p.m.—6:00 a.m.), fixed at 2.5 or 7.5 ach depending on the cross-ventilation efficacy (windows located in one facade or in two different facades according to UNI 10375: 2011).
- Natural ventilation: it is simulated as active all the day with 1.5 or 4 ach depending on the cross-ventilation efficacy (windows located in one facade or in two different facades).

The results (Fig. 4.10) show that the mechanical ventilation determines the highest operative temperatures. A mixed ventilation (CMV + free cooling) instead ensures the lowest operative temperatures only if the outdoor dry bulb temperature reaches high values (about 28°C), while it has similar behavior to the other ventilation patterns for typical summer conditions. The natural ventilation determines similar conditions to the hybrid one and results to be preferable in typical summer days. Extending the comfort evaluation to the entire summer season and calculating the hours in which the operative temperature falls out of the comfort range (table within the Fig. 4.10) it can be noticed that the natural ventilation ensures the lowest overheating hours. However, this solution should be adopted only on the hottest weeks of the summer (from July 1 to September 1) to avoid overcooled environments at night.

Moreover the better performance of the natural vented solution depends on the fixed air change per hour (based on the abovementioned standard) that during the day for the selected room is set at 4 ach for natural ventilation, while it assumes a very lower value of 0.5 in the mixed mode.

In summary the natural and hybrid ventilation result to be the most efficient cooling strategies especially when the outdoor temperature are not extremely hot. Inversely in very hot periods (and for extension in desert climates) CMV + free cooling should be preferred, to avoid heat entrance through ventilation.

4.3.5.3 Global evaluation
The overall evaluation from energy saving, comfort, and costs point of view of the various constructive solutions with different internal inertia by adopting different ventilations is shown in Fig. 4.11 and Table 4.7. The case with CMV without night free-cooling was excluded for its worst outcome.

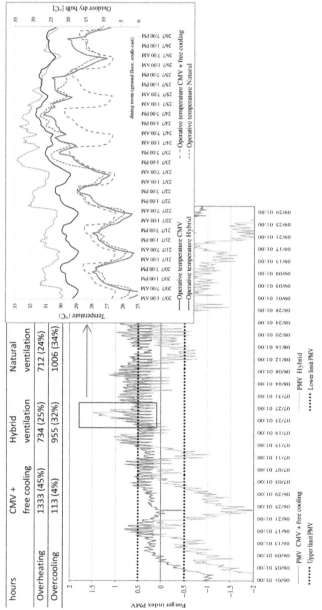

Figure 4.10 Summer comfort evaluation with different cooling strategies, southeast dining room on the ground floor: PMV comfort model (Method A-Annex F-EN 15251): operative temperatures for the hottest week. *From Super-insulated wooden envelopes in Mediterranean climate: summer overheating, thermal comfort optimization, environmental impact on an Italian case study, Energy Build. 138 (March 1, 2017) 716–732, ISSN 0378-7788.*

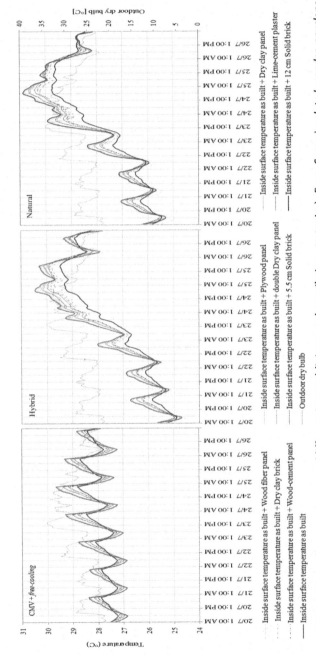

Figure 4.11 Inside surface temperatures (different internal linings and ventilation strategies). *From Super-insulated wooden envelopes in Mediterranean climate: summer overheating, thermal comfort optimization, environmental impact on an Italian case study, Energy Build. 138 (March 1, 2017) 716–732, ISSN 0378-7788.*

Table 4.7 Energy saving, indoor thermal comfort, and costs of the constructive solutions with different thermal inertia

Building consumptions: Winter = 6.47 kWh/m² year ± 0.07 (COP = 3.1); Summer = 4.03 kWh/m² year ± 0.08 (EER = 2.9)

	Low thermal capacity			Medium thermal capacity					High thermal capacity	
	Plaster-board	Wood fiber panel	Plywood panel	Dry clay panel	Dry clay brick	Double clay panel	Lime-cement plaster	Wood-cement panel	5.5 cm Solid brick	12 cm Solid brick
Summer discomfort (dining room—ground floor, southeast)										
CMV + free-cooling										
	1333	1357	1341	1330	1354	1327	1323	1341	1328	1360
	45.5%	46.3%	45.8%	45.4%	46.2%	45.3%	45.2%	45.8%	45.4%	46.4%
Hybrid										
	734	745	737	725	712	694	700	704	697	683
	25.1%	25.4%	25.2%	24.8%	24.3%	23.7%	23.9%	24.0%	23.8%	23.3%
Natural										
	712	704	706	708	692	695	703	694	692	670
	24.3%	24.0%	24.1%	24.2%	23.6%	23.7%	24.0%	23.7%	23.6%	22.9%
Building component cost (€/m²)										
	146	112	118	110	187	136	103	115	121	173
% Cost variation										
	—	−23	−19	−25	+28	−7	−29	−21	−17	+18

The graph highlights that the best solution is always the one with 12 cm thick solid bricks for its more favorable dynamic parameters combination. Moreover, the benefices of this solution are even more noticeable in very hot days and under dynamic ventilation strategies (hybrid or natural) that allow the mass cooling at night. The wood fiber panels show again the worst behavior.

The discomfort hours study (Table 4.7) confirms what has been obtained from the surface temperature analysis. However, the extension of simulated period to the entire hot season reveals that at its beginning and at its end, the 12 cm solid brick could cause overheating if combined with a conservative ventilation strategy. Natural and hybrid ventilations are the best solutions not only in the as built conditions but also at the varying of the inner linings. For these two strategies, the adoption of inner 12 cm solid bricks and the use of double clay panels are the best solutions, guaranteeing optimal summer comfort levels (23%−24% of summer discomfort hours). For CMV + free cooling instead the adoption of a solution with high decrementing attitude (12 cm solid brick) is not recommended since this type of cooling strategy has a daily low air exchange, thus more difficulty to dissipate the heat toward the outside. The consumption values are not detailed for each internal lining since the variation respect to the as built case falls within a reduced range ($\pm 0.07/0.08$ kWh/m^2 year). The costs study allows identifying one optimal solution, namely the introduction of two layers of dry clay panels.

In summary, for all ventilations adopted, the introduction of a double clay panel is a compromise with low costs and optimal thermal comfort levels. This lining is cheaper as well as structurally lighter and thinner (thus leaving more internal useful pavement area) than the 12 cm thick solid brick and reaches similar comfort levels. Even the lime-cement plaster could be a suitable solution but it is not recommended in dry techniques as wooden envelopes.

4.3.5.4 In other extremes climates

This section reports the results regarding the envelope in its initial configuration and two optimization solutions that achieved the best outcomes (12 cm thick solid bricks and double clay panels) at the varying of ventilation strategies in other extreme climates. The comfort conditions of a room (dining room at the ground floor) for the

Table 4.8 Incidence of the internal inertia on summer and winter comfort levels in various climate zones (PMV model and adaptive model)

Summer—PMV model % Hours of overheating	Ancona[a]	Palermo[a]	Cairo[a]
CMV + free-cooling	1333 (45.5%)	2187 (53.6%)	3022 (54.7%)
Hybrid	734 (25.1%)	1658 (40.6%)	2835 (51.4%)
Natural	712 (24.3%)	1527 (37.4%)	3126 (56.6%)
Double clay panel + CMV + free-cooling	1327 (45.3%)	2175 (53.3%)	3009 (54.5%)
Double clay panel + hybrid	694 (23.7%)	1638 (40.1%)	2838 (51.4%)
Double clay panel + natural	695 (23.7%)	1518 (37.2%)	3151 (57.1%)
12 cm solid brick + CMV + free-cooling	1360 (46.4%)	2183 (53.5%)	3045 (55.2%)
12 cm solid brick + hybrid	683 (23.3%)	1657 (40.6%)	2881 (52.2%)
12 cm solid brick + natural	670 (22.9%)	1519 (37.2%)	3162 (57.3%)
Summer—Adaptive model **% Hours of overheating**	**Ancona**[a]	**Palermo**[a]	**Cairo**[a]
Natural	134 (4.6%)	64 (1.6%)	1144 (20.7%)
Double clay panel + natural	121 (4.1%)	40 (1.0%)	1074 (19.5%)
12 cm solid brick + natural	88 (3.0%)	15 (0.4%)	1027 (18.6%)
Winter—PMV model **% Hours of overcooling**	**Ancona**[b]	**Bolzano**[b]	**London**[b]
Intermittent heating	906 (22.7%)	1536 (35.0%)	1346 (20.5%)
Double dry clay panel + Int. heat.	909 (22.8%)	1521 (34.6%)	1355 (20.6%)
12 cm solid brick + Int. heat.	905 (22.7%)	1506 (34.3%)	1372 (20.9%)

[a]Hot season length: Ancona, from June 1 to September 30; Palermo, from May 15 to October 31; and Cairo, from April 15 to November 30.
[b]Cold season length: Ancona, from November 1 to April 15; Bolzano, from October 15 to April 15; and London, from September 15 to June 15.

different climate zones were evaluated with both PMV and adaptive method (Table 4.8).

The comparison between different ventilation methods confirms that in mild climates such as temperate and semiarid zones (namely Ancona and Palermo) the most suitable solutions are either the adoption of hybrid ventilation (CMV + night natural ventilation) or daily natural ventilation, since in both climates the inner space could take advantage by the considerable seasonal and daily temperature range. Instead, the daily natural ventilation is not recommended in the constantly hot climate of Cairo.

In such climate, more "conservative" cooling solutions, e.g., mechanical ventilation and hybrid ventilation, are to be preferred.

The comparison between alternative inner linings highlights that the adoption of the double layer of clay panels, reduces (or leaves unchanged) the overheating hours respect to the as built solution for all climates, except for the extreme situation of natural vented buildings at Cairo. The adoption of a 12 cm thick bricks internal layer has a positive effect in Ancona and Palermo, while it has an opposite effect in an extremely hot climate (Cairo) since the high decrementing attitude of the thermal wave (low f value) determines the impossibility of the mass to be cooled at night.

Under natural ventilation conditions a comparison between PMV and adaptive model highlights that the former detects a percentage of discomfort very higher than that obtained with the latter model for the different range of comfort conditions adopted by the two models (see also discussion on this topic in Chapter 6: Experimental Methods, Analytic Explorations, and Model Reliability).

Fig. 4.12 reports the plotting of the operative temperatures obtained numerically for the dining room, with the various ventilation solutions in different zones, considering the same hot season length for all climates (June 1—September 30 for summer; November 1—April 15 for winter).

The best solutions in the temperate climate of Ancona (Italy) and in the semiarid climate of Palermo are confirmed to be both natural and hybrid ventilation, except for extreme summer week in which the very high external temperatures make preferable to adopt solution with low air exchange (Fig. 4.10, from July 23 to 25). The ventilation comparison in the desert climate of Cairo shows that the most conservative solution, namely the CMV with night free-cooling, is to prefer for the extreme external temperatures. This result slightly differs from the results regarding the discomfort hours in Table 4.8, in which hybrid ventilation was found to be preferable at Cairo. This is due to the fact that the sum of overheating hours is extended to the whole summer season, thus including periods with not so high temperatures, namely the beginning of the summer (from April 15 for Cairo) and the end of the hot period (to November 30). In such periods, CMV provides the highest operative temperatures.

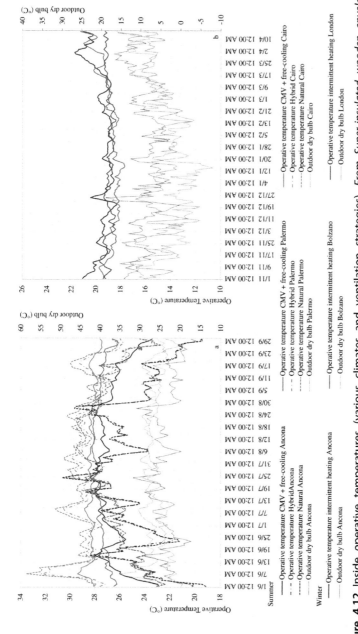

Figure 4.12 Inside operative temperatures (various climates and ventilation strategies). *From Super-insulated wooden envelopes in Mediterranean climate: summer overheating, thermal comfort optimization, environmental impact on an Italian case study, Energy Build. 138 (March 1, 2017) 716–732, ISSN 0378-7788.*

The winter behavior of the initial solution with intermittent heating was evaluated in other cold climates in Fig. 4.12. In the very cold alpine climate of Bolzano the operative indoor temperatures are lowered down to 1.5°C, respect to the values calculated for the milder climate of Ancona.

4.4 MASSIVE LOAD-BEARING WALL CONSTRUCTIONS

4.4.1 Problem description

Is it preferable choosing a two layered solution that maximizes the internal mass or combining in a single layer both storing and insulating functions thus strongly enhancing the decrementing attitude to the detriment of the inner inertia? Or adopting multilayered solutions with different layers characterized by highly specialized features? The present section tries to give an answer to these questions exploring different types of load-bearing walls and demonstrating that the selected energy efficient envelopes with different inertia present similar and very low consumptions whatever the system usage profiles but very different performances on indoor comfort levels and environmental impacts.

Consider the building components for the three techniques shown in Table 4.9. They will be adopted in three buildings with the same dimension and internal distribution. The buildings only differ for the opaque horizontal and vertical envelopes and the internal partitions that were designed with different construction techniques: masonry, wood–cement, and wood. The three techniques present unequal dynamic thermal characteristics but have the same, very low, thermal transmittance values of 0.26 W/m^2 K for the vertical walls and 0.32 W/m^2 K for the horizontal structures. The lightweight wooden solution presents high decrement factor value related to a low ability to damp the crossing thermal wave and it is characterized by inner layers (plasterboard and mineral insulation behind it) with very low conductivity (λ) and internal capacity (κ_1 of about 20 kJ/m^2 K) thus indicating a low attitude to transmit heat to the adjacent layers.

The other two techniques, masonry and wood–cement, are both massive but present external walls characterized by two different insulation-mass configurations. The former is stratified with very high internal mass

Table 4.9 Massive load bearing envelopes of a project designed by the Regional Public Housing Authority of Ancona province (E.R.A.P. Ancona)

Masonry	Wood-cement	Wood

External walls

Masonry

Lime-gy. plaster 1.5 cm

Hollow bricks 35 cm
Woodfiber insul. 10 cm
Lime-ce. plaster 2.5 cm

$U = 0.26$ W/(m²K)
$\kappa_1 = 42.95$ kJ/(m²K)
$f = 0.038$
$\Delta t = 20.07$ h

Wood-cement

Double plasterb. 2.5 cm

Wood-ce. block 38 cm
Lime-ce. plaster 2.5 cm

$U = 0.26$ W/(m²K)
$\kappa_1 = 37.45$ kJ/(m²K)
$f = 0.027$
$\Delta t = 16.98$ h

Wood

Double plasterb. 2.5 cm
Vertical air gap 1.5 cm
Mineralfiber insul. 5 cm
CLT (timber) 8.5 cm
Wood fiber insul. 8 cm
Lime-ce. plaster 2.5 cm
$U = 0.26$ W/(m²K)
$\kappa_1 = 20.36$ kJ/(m²K)
$f = 0.163$
$\Delta t = 11.49$ h

Floors over unheated space

Masonry

Ceramic tiles 1 cm
Screed coat 5 cm
XPS 7.5 cm
Ligh. concrete 7 cm
Concrete 4 cm
Brick floor slab 20 cm
Lime-gy. plaster 2 cm
$U = 0.32$ W/(m²K)
$\kappa_1 = 59.26$ kJ/(m²K)
$f = 0.097$

Wood-cement

Ceramic tiles 1 cm
Screed coat 5 cm
XPS 5 cm
Ligh.concrete 8 cm
Concrete 4 cm
Wood-ce. slab 20 cm
Lime-gy. plaster 2 cm
$U = 0.32$ W/(m²K)
$\kappa_1 = 58.55$ kJ/(m²K)
$f = 0.037$

Wood

Ceramic tiles 1 cm
Screed coat 5 cm
Mineral insul. 2 cm
Sand 5 cm
CLT (timber) 16.5 cm
Wood fiber insul. 4 cm
Plasterboard 2.5 cm
$U = 0.32$ W/(m²K)
$\kappa_1 = 56.12$ kJ/(m²K)
$f = 0.04$

Layers:	λ (W/(mK))	c (J/(kgK))	ρ (kg/m³)
Lime-gypsum plaster	0.700	1000	1400
Hollow clay brick	0.240	840	800
Wood fiber insul.	0.048	2100	230
Lime-cement plaster	0.900	1000	1800
Double Plasterboard	0.600	1000	750
Wood-cement block	0.107	1282	887
Mineral fiber insulat.	0.042	670	40
CLT (timber)	0.120	1600	500
Ceramic tiles	1.163	840	2300
Screed coat	0.930	1000	1800
XPS	0.035	1200	35
Ligh. concrete	0.175	880	500
Concrete	0.930	1000	1800
Brick floor slab	0.940	840	1800
Wood-cement floor slab	0.236	2100	500

Source: From F. Stazi, E. Tomassoni, C. Bonfigli, C. Di Perna, Energy, comfort and environmental assessment of different building envelope techniques in a Mediterranean climate with a hot dry summer, Appl. Energy 134 (December 1, 2014) 176–196, ISSN 0306-2619.

(κ_1 of about 43 kJ/m^2 K), while in the latter solution the inner layer is composed of a plastered wooden-cement mixture with lower mass and as a consequence a lower storage ability (κ_1 value of 37 kJ/m^{2}K). Moreover, the wood-cement blocks have a very low λ value respect to the masonry hollow blocks thus presenting lower capacity to conduct the heat. The wood-cement solution determines a very low decrement factor that could impede the mass cooling through heat losing toward the nocturnal sky.

Nowadays there is a growing awareness that all these highly insulated solutions could act as thermal barriers respect to the external temperature variations, losing the dynamic interaction with the specific climate typical of massive not insulated traditional walls. Various authors demonstrated that this problem could be reduced adopting an external position for the insulation layer respect to the massive one, but few studies were performed on the preferable characteristics that this inner massive layer should have [16,17]. These studies are generally focused on a specific climate and on a prevailing issue but the mass could have a contrasting impact on different aspects: environmental burdens [18,19], energy saving, or cost effectiveness, the last two cases depending on climates or usage patterns [20,21].

Some authors [18] contributed to fill this gap by analyzing the mass effect through the combined evaluation of two different approaches such as energy saving and life cycle primary energy balance. The authors demonstrate that the environmental advantages of a lightweight wooden envelope outweigh the energy saving benefits of thermal mass. However, the study is realized in an extremely cold climate and with an internal fixed temperature (without evaluating different usage patterns), so that the mass benefits could not be fully appreciated and the summer comfort levels were not considered.

An open question is then the impact of various mass and superinsulation configurations on all the abovementioned conflicting aspects in various hot and cold climates and for different usage patterns. The following section tries to contribute to the debate through analytical simulations on the three techniques.

4.4.2 Comfort

In summer the low outdoor—indoor temperature gradient combined with the presence of a high insulation thickness and the low f values (especially for the masonry and wood-cement techniques) determine that the flux does not entirely cross the wall and the greatest amount of heat gains comes from the internal side (mainly for greenhouse effect through the glazed surfaces). Therefore the inner layer (just adjacent to the room) assumes a greater importance than the other layers behind it to enhance the dynamic interaction with the internal overheated environment. It is important to adopt a highly capacitive and highly conductive inner layer, two characteristics maximized in the inner materials of the masonry technique. This technique thanks to the favorable thermal dynamic parameters is able to reach the best comfort levels, as demonstrated in Fig. 4.13 that identifies different phases.

A first phase (spring), in which the two massive walls have very similar results, while the lightweight one is characterized by greater maximum values (1.5−2°C). With the seasonal increasing of the external temperature, the difference between massive and lightweight solutions is more pronounced, underlining the overheating attitude of the wooden envelope.

A second phase in summer characterized by elevated outside temperatures in which the overheating of the interior space of the low capacity building is confirmed. However, as highlighted elsewhere (Section 2.5.2) at the end of the hot period there is a trend inversion since the lowering of the external temperature suddenly affects the performance of the not massive solution with a rapid reduction of the operative temperatures, while the massive walls show a more conservative behavior, maintaining higher operative temperatures until autumn. The day of trend inversion is highlighted in the graph through a *dotted line*.

In autumn (September−October) an inverted behavior is thus shown. In such period, the external air temperature causes the internal reduction of the operative temperatures especially in the lightweight building. So the latter technique approaches the beginning of the cold season in a disadvantaged condition. The wood–cement solution shows the best trend for its lowest decrement factor that enhances the wall attitude to conserve the inner stored heat.

The analysis of the discomfort hours with the adaptive method in the whole summer (Table 4.10) could be misleading if not accompanied by the values subdivided in the two phases, before and after the trend inversion. In the first part of the summer the massive solutions record the lowest discomfort hours for overheating, while in the second period they approach the winter with the highest internal temperatures (with the best performance for the wood-cement solution). The discomfort hours in

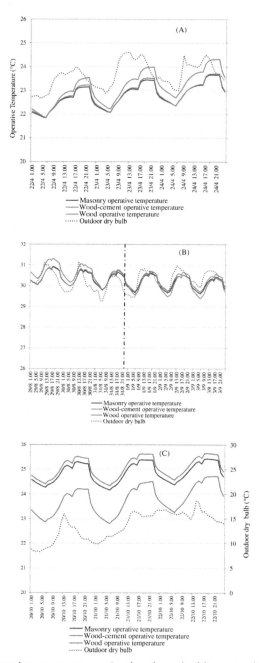

Figure 4.13 Operative temperatures in the three buildings in different seasons. (A) Spring, (B) summer, and (C) autumn.

Table 4.10 Extensive comparison of the three buildings: summer comfort and cooling-heating consumptions
Construction techniques/Ancona

	Summer (June 1–September 30) Hours of overheating % Hours on total Adaptive model	PMV model	Cooling consumptions [b] (kWh prim./m² year)	Heating consumptions [b] (kWh prim./m² year)
Masonry	643 (285 + 358)[a] 22.0%	2454 (1811 + 643)[a] 83.8%	5.86	19.43
Wood-cement	745 (328 + 417)[a] 25.4%	2471 (1817 + 654)[a] 84.4%	5.93	19.37
Wood	695 (473 + 222)[a] 23.7%	2492 (1893 + 599)[a] 85.1%	6.03	19.54

[a]The values are obtained as the sum of discomfort hours in the first period (until 1st September) and of the second period (from 1st September). The 1st September is the day where there is a trend inversion between the three techniques for Ancona (see Section 2.5.2).
[b]Intermittent operation, 2 sections (5:00 a.m.–11:00 a.m., 3:00 p.m.–11:00 p.m.).

summer were also calculated with PMV model to compare the results obtained with the two models despite the Standard EN 15251 suggests associating this type of calculation to a conditioned use of the apartment (see discussion in Section 6.5.2). The PMV model detects a substantial percentage of discomfort (84%−85%), very higher than that obtained with adaptive comfort model (between 22% and 25%).

4.4.3 Consumptions

The dynamic energy analysis for the city of Ancona (Table 4.10) shows a very similar behavior of the three strategies, with a difference of about 1% in winter and 2.5% in summer between the best and the worst solution.

In winter the great differences between external temperature (generally around 5−10°C even for the mild climate of Ancona) and internal temperature (set point at 20°C) creates an outgoing flux that involves the entire wall stratigraphy. As a consequence the entire envelope mass contributes with its damping effect, which becomes the prominent feature that affects the winter performance (when the transmittance is fixed). The wood-cement solution for its lowest decrement factor maximizes this behavior, thus providing the lowest consumptions.

In summer, the masonry wall results to have the best performance due its higher internal inertia (κ_1 value). In that period the lower difference between internal and external temperature than those recorded in winter determines that both external and internal surfaces will tend toward average temperatures thus reducing the crossing heat flux and giving importance to the wall's layer adjacent to the inner environment.

The wood is the worst solution for all year round, even if with small differences respect to the other two solutions since it has the worst dynamic thermal parameters: the lowest internal areal heat capacity (and as a consequence the lowest inner inertia) and the lowest damping effect on the incoming wave.

The study of the thermal bridges was done with the aid of *Therm* software (Table 4.11). The location of each thermal bridge is clearly visible in Fig. 2.10. For all the techniques, the main problem is the connection between the glazed facade portion and the opaque surface. Moreover, especially for the masonry technique great attention should be paid at the connection between the floor lying above the unheated garage and the

Table 4.11 Thermal bridges
Thermal bridges, ISO 14683: $\Psi = L^{2D} - \Sigma U_j l_i$

Type of thermal bridge[a]/linear thermal transmittance[b]	Masonry Ψ_i (W/ mK)	Wood-cement Ψ_i (W/ mK)	Wood Ψ_i (W/ mK)
C1—corner of external brick walls	0.13	0.07	0.05
C2—corner of external plastered walls	0.13	0.07	0.05
C3—corner of external walls	−0.17	0.07	0.06
IW1—internal load-bearing wall/external wall	0.05	0.19	0.05
IW2—internal load-bearing wall/corner of external walls	−0.04	0.10	−0.05
IW3—corner of unheated space walls/external plastered wall	0.11	0.20	0.05
IF1—intermediate floor/external brick wall	0.09	0.18	0.09
IF2—intermediate floor/external plastered wall	0.09	0.19	0.08
GF—ground floor/external plastered wall/roof of unheated space	0.20	0.07	0.16
R1—roof/external plastered wall	0.01	0.07	0.03
R2—roof/beam/external plastered wall	0.13	0.16	0.11
R3—beam/external brick wall	0.06	0.02	0.00
R4—roof/beam	0.04	0.04	0.04
R5—roof/external wall	0.07	0.07	0.07
W1—window/external brick wall	0.24	0.22	0.20
W2—window/external plastered wall	0.24	0.22	0.19
Total: $\Sigma\Psi_i l_i$ (W/K)	112	139	101

[a]C, corners; GF, suspended ground floors; IF, intermediate floors; IW, internal walls; R, roofs; W, windows.
[b]For the calculation the internal dimensions measured between the finished internal faces of each room in a building have been considered. The thermal coupling coefficient L^{2D} (W/m K) is determined through *Therm* software.
Source: From F. Stazi, E. Tomassoni, C. Bonfigli, C. Di Perna, Energy, comfort and environmental assessment of different building envelope techniques in a Mediterranean climate with a hot dry summer, Appl. Energy 134 (December 1, 2014) 176–196, ISSN 0306-2619.

vertical external walls. Instead, for the wood-cement solution a considerable heat loss occurs at the intermediate floors to vertical walls corners. As expected the wooden solution characterized by a material with low thermal conductance has the best behavior regarding the thermal bridges.

4.4.4 Environmental sustainability

The study of the impacts distribution in the various life cycle phases (Fig. 4.14) with Eco-indicator 99, CED, and EPS 2000 methods (*Weighting* analysis) shows that the greatest total damage of the three solutions is mainly due to the initial "construction" phase according to all methods. This "construction" phase is more impacting for wood in Eco-indicator and CED, while for wood-cement in EPS 2000. This is because the first two methods give more relevance to ecosystem and renewable sources, while the third is more focused on human health. The "transport" phase has the great burdens for the wooden solution since it is not a locally available material with higher fossil fuels to carry it to the building site. The wooden solution has also the greatest impact in "maintenance" phase with CED method because of renewable sources use, while the wood-cement results to have the greatest maintenance burdens by adopting the other two methods. As results from energy analysis, there are not relevant differences between the three techniques for "winter use" and "summer use" phases with the wood worst behavior. Concerning the "end-of-life," the wood technique has the greatest benefit for the energy recovered from the wood incineration and for the greater credits of recycling resulting from the subtraction from the system of the environmental burdens associated with manufacturing of the substituted product.

The graph in Fig. 4.15 reports the evaluation with Eco-indicator 99 focusing on the construction phase that came out to be the biggest weight on global LCA disregarding the materials with low impact and those that, even if with high impacts (such as laminated wood and PVC), are in common to all techniques. The graph shows that the highest impact is due to the basic materials characterizing the three envelopes and the relative structures:

- For masonry: bricks for horizontal and vertical envelope (*Hollow clay brick* and *Brick floor slab*) and concrete used mainly for horizontal structures (*Concrete* and *Reinforced concrete*) have the major impacts, respectively, on "Resources" (use of *Fossil fuel* to fire brick kilns) and "Human Health" (respiratory diseases linked to cement production process). The insulation materials have smaller impacts and the *wood fiber insulation* is the only material with a negative impact, so its use avoids environmental damage rather than produces it. The finishes have very low impact.

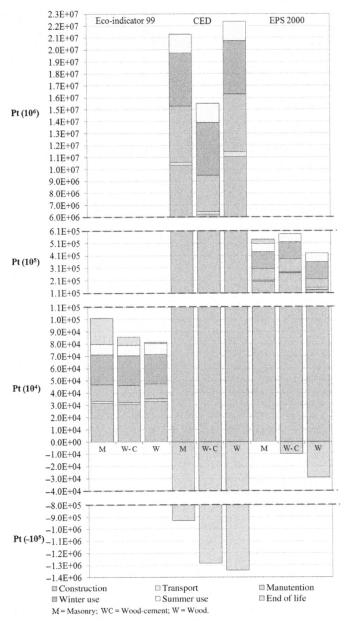

Figure 4.14 LCIA: life cycle phases results of three buildings during 75-year life span with Eco-indicator 99, CED, and EPS 2000 methods. *From F. Stazi, E. Tomassoni, C. Bonfigli, C. Di Perna, Energy, comfort and environmental assessment of different building envelope techniques in a Mediterranean climate with a hot dry summer, Appl. Energy 134 (December 1, 2014) 176–196, ISSN 0306-2619.*

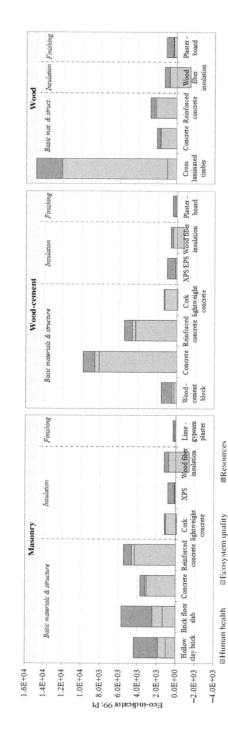

Figure 4.15 Construction phase results of three buildings at year 0 with Eco-indicator 99 method. *From F. Stazi, E. Tomassoni, C. Bonfigli, C. Di Perna, Energy, comfort and environmental assessment of different building envelope techniques in a Mediterranean climate with a hot dry summer, Appl. Energy 134 (December 1, 2014) 176–196, ISSN 0306-2619.*

- For wood-cement: *Wood-cement blocks* (basic material characterizing this technique) and concrete used for the filling of the envelope cavities and for horizontal structures (*Concrete* and *Reinforced concrete*) have the major impacts, respectively, on "Resources" (*Fossil fuel* used for forming solid blocks and for the milling process) and "Human Health" (for the abovementioned reason). In addition, among the insulation material the highest impact is due to the EPS inserted in the blocks. This synthetic insulation, even if employed in very smaller quantities than the natural ones, present higher impacts for the industrial sintering process (mainly affecting fossil fuel "Resources").
- For wood: *CLT* has the greatest burden mainly influencing "Ecosystem Quality" category (for the high *Land use* necessary for the planting of trees). In this technique, the insulations have low impacts being completely of natural origin. The lower impact for cement and concrete is because of major use of dry techniques for this solution.

The LCA during 75-year life span shows that the three buildings present different impacts depending on the method used and the category analyzed (Table 4.12). All the results were expressed according to the Characterization phase (category indicators) and only the overall results of each method are reported in relation to the Weighting analysis (Pt). Nevertheless, for each category the multipliers are reported to appreciate the relative importance of the same category.

The study with Eco-indicator 99 method shows that the wooden building has the lowest total environmental impact while the masonry presents the greatest one. This is because in "Human Health" category the wooden solution is less impacting even if in "Ecosystem Quality" it produces a great damage. For the first category ("Human Health") the low impact of the wooden envelope is due to the use of ecological materials that reduce the release of carcinogenic chemicals and inorganic particles in environments that are harmful for toxicological risk and respiratory health (Carcinogens, Respiratory inorganic) and do not affect the atmospheric composition (Climate change). For the second category ("Ecosystem Quality") the great damage of the lightweight solution is mainly due to the transformation of land area for human activities (Land use) since the wood is inevitably extracted from forests and green areas. For the third category ("Resources") the wooden technique has an intermediate behavior mainly due to mid burdens in Fossil fuels subcategory. The worst behavior between the three techniques is that of masonry, mostly caused by the "Resources" category for highly impacting Fossil Fuel required to fire brick kilns.

Table 4.12 LCIA results of the three buildings during 75-year life span with Eco-indicator, CED, EPS 2000, and IPCC 2001 GWP methods

Method/construction techniques	Characterization / Unit	Conversion to Pt	Masonry	Wood-cement	Wood
Eco-indicator 99 (H)—hierarchist perspective					
Human health	**DALY**	$\times 65.1\ DALY^{-1}\times 300Pt$	**1.76E + 00**	**1.60E + 00**	**9.52E − 01**
Carcinogens	DALY		9.54E − 01	5.29E − 01	4.06E − 01
Respiratory organics	DALY		9.40E − 04	8.27E − 04	1.01E − 03
Respiratory inorganics	DALY		7.62E − 01	1.01E + 00	5.84E − 01
Climate change	DALY		3.72E − 02	6.13E − 02	−4.05E − 02
Radiation	DALY		1.47E − 03	1.37E − 03	1.98E − 03
Ozone layer	DALY		1.77E − 04	1.75E − 04	1.79E − 04
Ecosystem quality	**PDF m² year**	$\times 1.95E-4\ P.m^2/year\times 400Pt$ $(=0.1\times PAF\ m^2\ year)$	**2.76E + 05**	**1.74E + 05**	**2.67E + 05**
Ecotoxicity	PDF m² year		1.77E + 05	9.79E + 04	6.70E + 04
Acidification/eutrophication	PDF m² year		1.01E + 04	9.97E + 03	1.08E + 04
Land use	PDF m² year		8.88E + 04	6.65E + 04	1.89E + 05
Resources	**MJ surplus**	$\times 1.19E-4\ MJ/s\times 300Pt$	**1.28E + 06**	**1.15E + 06**	**1.20E + 06**
Minerals	MJ surplus		1.31E + 04	1.06E + 04	1.18E + 04
Fossil fuels	MJ surplus		1.27E + 06	1.14E + 06	1.19E + 06
Total	**Pt**		**1.02E + 05**	**8.59E + 04**	**8.24E + 04**
Cumulative energy demand—CED					
Not renewable sources	**MJ-Eq**	$\times 1$	**1.00E + 07**	**9.04E + 06**	**9.87E + 06**
Fossil	MJ-Eq		9.47E + 06	8.52E + 06	9.07E + 06
Nuclear	MJ-Eq		5.63E + 05	5.17E + 05	8.04E + 05
Renewable sources	**MJ-Eq**	$\times 1$	**1.03E + 07**	**5.16E + 06**	**1.12E + 07**
Biomass	MJ-Eq		1.00E + 07	4.84E + 06	1.08E + 07
Wind, solar, geothermal	MJ-Eq		8.21E + 03	7.11E + 03	1.56E + 04
Water	MJ-Eq		3.24E + 05	3.18E + 05	3.55E + 05
Total	**Pt**		**2.04E + 07**	**1.42E + 07**	**2.10E + 07**

Environmental priority strategy—EPS 2000

Human health	Person year		1.91E+01	2.16E+01	1.73E+01
Life expectancy	Person year	×85,000 ELU/P.Year ×1Pt	2.30E+00	3.33E+00	1.16E+00
Severe morbidity	Person year	×100,000 ELU/P.Year ×1Pt	8.24E-02	1.09E-01	-6.11E-02
Morbidity	Person year	×10,000 ELU/P.Year ×1Pt	1.82E-01	2.41E-01	-8.55E-02
Severe nuisance	Person Year	×10,000 ELU/P.Year ×1Pt	2.79E-01	2.80E-01	2.88E-01
Nuisance	Person year	×100 ELU/P.Year ×1Pt	1.63E+01	1.76E+01	1.60E+01
Ecosystem production capacity	**kg**		**3.58E+04**	**2.50E+04**	**4.19E+04**
Crop growth capacity	kg	×0.15 ELU/kg ×1Pt	2.17E+03	1.96E+03	1.45E+03
Wood growth capacity	kg	×0.04 ELU/kg ×1Pt	1.66E+03	-8.81E+03	8.23E+03
Fish and meat production	kg	×1 ELU/kg ×1Pt	-5.70E+01	-5.32E+01	-5.61E+01
Soil acidification	H + eq.	×0.01 ELU/H + eq. ×1Pt	3.86E+03	3.81E+03	4.11E+03
Prod. cap. irrigation water	kg	×0.003 ELU/kg ×1Pt	1.41E+04	1.41E+04	1.41E+04
Prod. cap. drinking water	Kg	×0.03 ELU/kg ×1Pt	1.41E+04	1.41E+04	1.41E+04
Abiotic stock resources	**ELU**		**3.28E+05**	**2.73E+05**	**2.99E+05**
Depletion of reserves	ELU	×1 ELU/ELU ×1Pt	3.28E+05	2.73E+05	2.99E+05
Biodiversity	**NEX**		**1.17E-08**	**1.18E-08**	**1.39E-08**
Species extinction	NEX	×1.1E11 ELU/NEX ×1Pt	1.17E-08	1.18E-08	1.39E-08
Total	**Pt**		**5.41E+05**	**5.75E+05**	**3.98E+05**

Global potential warming—IPCC 2001 GWP

IPCC GWP 100a	kg CO_2-Eq.		2.12E+05	3.11E+05	-1.80E+05

Source: From F. Stazi, E. Tomassoni, C. Bonfigli, C. Di Perna, Energy, comfort and environmental assessment of different building envelope techniques in a Mediterranean climate with a hot dry summer, Appl. Energy 134 (December 1, 2014) 176–196, ISSN 0306-2619.

The study with CED method shows instead that the wood-cement has the lowest total burdens and the wood (unlike the previous analysis) presents the greatest one even if due to a major exploitation of "Renewable Sources," mainly from Biomass. The wood-cement has the smallest impact in both categories and in particular as regards the use of renewable sources.

The study with EPS 2000 method shows that the total lowest impact is obtained from the wooden building while the greatest one from the wood-cement solution. The wood has the smallest impact in "Human Health" and the greatest damage in "Ecosystem Production Capacity," mainly because of a reduced Wood Growth Capacity, and in "Biodiversity," affecting the ecosystem Species Extinction. The highest total damage of wood-cement is caused by "Human Health" category due to dust and toxic emissions released during the preparation of the concrete, as resulted in Eco-indicator 99.

The study with IPCC 2001 GWP demonstrates that the wooden building causes the lower carbon dioxide emissions reaching even a value of negative impact (net absorption of emissions). It results to be creditor in terms of CO_2-Eq because the balance takes into account the CO_2 stored by wood during its life cycle. The wood-cement solution causes the greatest CO_2 concentration in the atmosphere for the use of concrete, contributing to climate change. The CO_2 emissions of masonry record intermediate values, mainly due to the fossil fuel energy required to fire brick kilns.

4.4.5 Optimized solutions

4.4.5.1 Increase of the internal inertia and ventilation strategies

The adoption of an inertial massive lining in the worst (wooden) solution makes it possible to achieve better comfort levels even in this technique, especially if combined with effective cross-ventilation.

Consider the thermal properties of the linings in Table 4.13. The adoption of a wood-cement panel in the internal side of the wooden solution allows reaching optimal dynamic parameters. It results in a surface temperatures lowering of 1.5°C (Fig. 4.16). The introduction of effective ventilation at night (2.5 ach) strongly reduces the differences between the techniques but further reduces the temperature values.

The two massive linings improve the attitude of the wall to decrement the thermal wave (through the reduction of f value) and strongly enhance the interaction with the internal environment through the increase of the

Table 4.13 Inner lining alternatives

	Walls			Floors		
	Plaster board[a]	Dry clay panel	Wood-cement	Ceramic tiles[a]	Dry clay panel	Wood-cement
Internal lining properties						
Thick. (cm)	2.5	2.2	4	1	2.2	2.8
λ (W/(m K))	0.6	0.47	0.26	1.163	0.47	0.26
c (J/(kg K))	1000	1000	1880	840	1000	1880
ρ (kg/m^3)	750	1300	1400	2300	1300	1400
Building components thermal properties						
U (W/(m^2 K))	0.26	0.26	0.26	0.32	0.32	0.32
κ_1 (kJ/(m^2 K))	20.36	32.50	61.79	56.12	56.11	56.09
f (−)	0.163	0.157	0.087	0.04	0.039	0.032
Δt (h)	11.49	12.06	15.40	18.31	18.57	20.57

[a]Linings adopted in the as built condition.

internal inertia (high κ_1 values). The wood-cement inner lining shows the best thermal parameters thus achieving the best comfort conditions. However, the study of the discomfort hours (Table 4.14) in the entire summer season could create some confusion since the CLT technique with both the two massive lining seems to have a worst behavior. Only the analysis of the values detailed before and after the trend inversion for the technologies (see Section 2.5.2) explains the data. The overheated hours for the central summer period (before trend inversion) decrease from 473 to 458 and 444, respectively, with the inner clay panel and the inner wood-cement panel. The inverse trend is shown at the end of the hot period. So the solution with wood-cement panel begins the cold season in a more advantaged condition.

In summary, the wood-cement panel has the best behavior thanks to its favorable dynamic parameters.

An experimental evidence of the benefice of adopting a massive lining in the internal side of a cross laminated timber envelope is reported in Appendix A, case A.1.2.

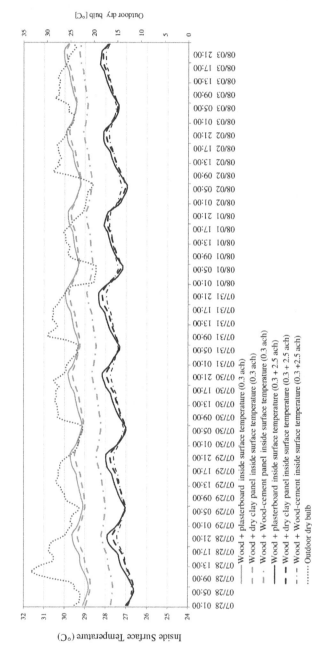

Figure 4.16 Summer week with high temperatures: inside surface temperatures of a wall in the living room, southern exposure with different internal lining and different profiles of natural ventilations.

Table 4.14 Numerical simulations

Adaptive model Summer (June 1– September 30)	Wood + platerboard	Wood + dry clay panel	Wood + wood- cement panel
Hours of overheating	695 (473 + 222)[a]	699 (458 + 241)	773 (444 + 329)
% Hours on total	23.7	23.8	26.3

Summer discomfort hours for cross-laminated timber technique with three different inner linings, city of Ancona.
The 1st September is the day where there is a trend inversion between the three techniques for Ancona.
[a]The values are obtained as the sum of discomfort hours in the first period (until 1st September) and of the second period (from September 1).

4.4.5.2 In other extremes climates

This variable was deepened in Section 2.6 for the same three technologies, masonry, wood-cement, and wood, named, respectively, W4, W5, and W6. Masonry and wood-cement solution have very similar performance also in extreme climates; the former is slightly preferable in extremely hot climates, while the latter is to prefer in cold ones.

4.5 DESIGN PATTERNS

The comfort monitoring demonstrated the presence of considerable indoor overheating in light-framed constructions, occurring mainly at the upper levels. Indeed the envelope behaves as a thermal barrier for its extremely low thermal transmittance. The heat dispersions occur only through the ground floor determining a noticeable air stratification especially in the central hours of the day. The numerical research demonstrated that such problem could be reduced through the introduction of an internal massive layer, such as 12 cm thick brick counter-walls or two layers of dry clay panels. An experimental evidence demonstrated that a 5 cm thick layer of solid bricks, improves the wall behavior without reducing the internal floored space. The use of hybrid ventilation is recommended. It is clearly evident that the hyperinsulation of new building techniques has strongly reduced the impact of one type of energy (natural gas used for winter heating) increasing on the other hand the use of another type of more environmental impacting energy (electricity for summer cooling). Among the materials adopted the thick insulations determine the highest environmental impact even if guarantee low winter consumptions, highlighting the conflicting incidence between the energy and environmental aspects.

Among the massive load-bearing walls, the optimal solution varies according to the considered aspect among the following: winter and summer consumptions, comfort conditions in winter, in intermediate seasons and in summer, environmental impacts, and cost effectiveness. The choice should depend on the specific case study and climatic location. In low-energy buildings, the small range between consumptions of the worst and best solution determines that the comfort in unconditioned period and environmental aspects prevail on energy issue. Between these two aspects, the former becomes a priority in climates with hot dry summer for the presence of extensive periods with high temperatures. In such climates, the thick insulations should be combined with high internal masses (masonry) capable to interact dynamically with the internal environment. The optimal configurations are those with high internal inertia and high (but not excessively) attenuating attitude. The adoption of a wooden load-bearing envelope, even if not convenient with regard to comfort aspects for the overheating phenomena, is preferable for lower exploitation of environmental resources and economic convenience. Nevertheless, to enhance its summer behavior, passive cooling techniques (such as the introduction of inner massive finishing such as wood-cement panel or a double layer of clay panels and the cross-ventilation) could be adopted.

REFERENCES

[1] L. Wang, J. Gwilliam, P. Jones, Case study of zero energy house design in UK, Energy Build. 41 (2009) 1215−1222.
[2] A. Lenoir, F. Garde, E. Ottenwelter, A. Bornarel, E. Wurtz, Net zero energy building in France: from design studies to energy monitoring. A state of the art review. Towards Net Zero Energy Solar, Buildings, IEA SHC Task 40 and ECBCS Annex (2008) 52.
[3] A.J. Marszal, P. Heiselberg, J.S. Bourrelle, E. Musall, K. Voss, I. Sartori, et al., Zero energy building—a review of definitions and calculation methodologies, Energy Build. 43 (2011) 971−979.
[4] F. Chlela, A. Husaunndee, C. Inard, P. Riederer, A new methodology for the design of low energy buildings, Energy Build. 41 (2009) 982−990.
[5] S. Thiers, B. Peuportier, Thermal and environmental assessment of a passive building equipped with an earth-to-air heat exchanger in France, Solar Energy 82 (2008) 820−831.
[6] A. Audenaert, S.H. De Cleyn, B. Vankerckhove, Economic analysis of passive houses and low-energy houses compared with standard houses, Energy Policy 36 (2008) 47−55.

[7] A. Ferrante, M.T. Cascella, Zero Energy balance and zero on-site CO_2 emission housing development in the Mediterranean climate, Energy Build. 43 (2011) 2002–2010.

[8] A. Giovanardi, A. Troi, W. Sparber, P. Baggio, Dynamic simulation of a passive house in different locations in Italy, in: PLEA 2008—25th International Conference on Passive and Low Energy Architecture. Dublin, 2008.

[9] P. Hernandez, P. Kenny, From net energy to zero energy buildings: Defining life cycle zero energy buildings (LC-ZEB), Energy Build. 42 (2010) 815–821.

[10] A.D. Kellenberger, H. Althaus, Relevance of simplifications in LCA of building components, Build. Environ. 44 (2009) 818–825.

[11] L. Zhu, R. Hurt, D. Correia, R. Boehm, Detailed energy saving performance analyses on thermal mass walls demonstrated in a zero energy house, Energy Build. 41 (2009) 303–310.

[12] D.S. Parker, Very low energy homes in the United States: perspectives on performance from measured data, Energy Build. 41 (2009) 512–520.

[13] J. Schnieders, A. Hermelink, CEPHEUS results: measurements and occupant's satisfaction provide evidence for Passive Houses being an option for sustainable building, Energy Policy 34 (2006) 151–171.

[14] S.F. Larsen, C. Filippìn, S. Gonzàlez, Study of the energy consumption of a massive free-running building in the Argentinean northwest through monitoring and thermal simulation, Energy Build. 47 (2012) 341–352.

[15] E. Krüger, B. Givoni, Thermal monitoring and indoor temperature predictions in a passive solar building in an arid environment, Build. Environ. 43 (2008) 1792–1804.

[16] C. Di Perna, F. Stazi, A. Ursini Casalena, M. D'Orazio, Influence of the internal inertia of the building envelope on summertime comfort in buildings with high internal heat loads, Energy Build. 43 (2011) 200–206.

[17] N. Aste, A. Angelotti, M. Buzzetti, The influence of the external walls thermal inertia on the energy performance of well insulated buildings, Energy Build. 41 (2009) 1181–1187.

[18] A. Dodoo, L. Gustavsson, R. Sathre, Effect of thermal mass on life cycle primary energy balances of a concrete and a wood-frame building, Appl. Energy 92 (2012) 462–472.

[19] R. Broun, G.F. Menzies, Life cycle energy and environmental analysis of partition wall systems in the UK, Proc. Eng. 21 (2011) 864–873.

[20] S. Chiraratananon, V.D. Hien, Thermal performance and cost effectiveness of massive walls under thai climate, Energy Build. 43 (2011) 1655–1662.

[21] B.M. Lj, D.L. Loveday, The influence on building thermal behaviour of the insulation/masonry distribution in a three-layered construction, Energy Build. 26 (1997) 153–157.

CHAPTER FIVE

Passive Envelopes

5.1 INTRODUCTION

The new European Directives [1,2] have stressed the importance of adopting strategies that contribute to improving the thermal performance of buildings during the summertime and encourage their use. In this regard, ventilated facades and passive solar walls are of considerable interest since they help to enhance the internal climate conditions and determine a reduction in the use of cooling energy during the summer.

However, low thermal resistance necessarily characterizes both ventilated facades and solar systems, since they work as thermal dissipating building elements in summer and, in the latter case as sun capturing systems. Therefore they present some drawbacks connected with the recently adopted superinsulated model in which thick insulations separate the mass from the vented cavity.

This chapter reports experimental measures on two types of recently built ventilated walls with different insulation-mass positions and on solar vented and unvented passive walls. The survey also included different wall features, such as chimney heights, exposure, and presence of shadings or overhangs. Moreover, the study addresses, through dynamic simulations, the effect of increasing levels of insulation in both ventilated facades and passive walls.

Table 5.1 and Fig. 5.1 show the sections of the passive system studied and the introduction of a thick insulation layer on a typical vented wall.

5.2 VENTILATED WALLS

5.2.1 Problem description

In hot-summer Mediterranean climates, the ventilated facades (especially those with opaque facing) have been increasingly used because they

Thermal Inertia in Energy Efficient Building Envelopes.
DOI: http://dx.doi.org/10.1016/B978-0-12-813970-7.00005-4

Table 5.1 Ventilated facades and Trombe walls

Ventilate facades	Solar walls[a]/Trombe walls
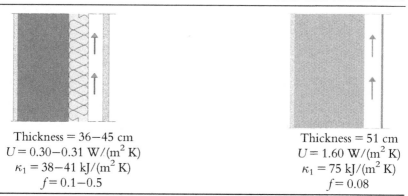	
Thickness = 36−45 cm	Thickness = 51 cm
$U = 0.30-0.31$ W/(m^2 K)	$U = 1.60$ W/(m^2 K)
$\kappa_1 = 38-41$ kJ/(m^2 K)	$\kappa_1 = 75$ kJ/(m^2 K)
$f = 0.1-0.5$	$f = 0.08$

[a]Solar walls are the same system but without the ventilation in the cavity.

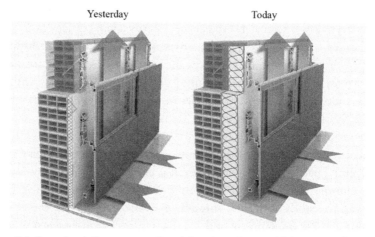

Figure 5.1 Superinsulation on ventilated facades. *Source: Elaboration of an image from edilceramiche.net.*

resolve the problem of the durability of the outer finishes of the external insulation layer, mainly caused by cracking because of aggressive solar radiation and rainfall, without the drawback of summertime overheating which is necessarily caused by double skin glazed facades. Ventilated facades with opaque cladding are generally characterized by the presence of one continuous insulation material lying next to the internal mass and an external protective cladding fastened to the wall using mechanical systems. A naturally ventilated channel is thus created.

The adoption of thick insulations interposed between the air cavity and the inner masses would strongly reduce the benefices of summer ventilation (impeding the heat to be removed from the inner mass) and as a consequence would cut the contribution to cooling energy saving given by such technologies. The final behavior of the wall will depend on the layers position and types.

Some characteristics influence the thermal and energy performance of ventilated facades:

1. The width and height of the ventilation channel [3,4]: a sensible cooling effect is obtained when the chimney width is wider than 7 cm, it is optimal for values around 10−15 cm, and becomes stable for higher dimensions. The chimney width affects the natural convection and air movement inside the cavity, since its increase determines the enhancing of the stack effect and the consequent increasing of airflow rate for the reduction of wall frictional resistance. Decreasing the ratio between width and height of the channel, e.g., increasing the chimney height with fixed gap width, results in an airflow rate increase.

2. The factors connected with the site (solar radiation, wind, and exposure) which determine the local microclimate [5−7].

3. The type of external cladding and the characteristics of the materials that are laid adjacent to the channel [8,9]. The cladding may be made of a thin metal (e.g., zinc-titanium) supported by a wooden plank or a thicker solid material (e.g., brick, ceramics, cement) [10] and may be permeable (with open joints) [11,12] or airtight (closed joints). Ventilated walls with integrated PV panels are increasingly used [13,14]. Geometry and mean roughness of the surfaces within the channel affect the uniformity and continuity of the heat flux, the surface temperature variation along the channel and the pressure losses. Some author stressed that the introduction of massive layers within the channel could be not beneficial in winter for the airflow rate increase [15]. The presence of open joints determines a greatest influence of the wind on airflows.

Actually, there is a lack of experimental studies aimed at quantifying the importance of the features specified above, only analytically investigated by some author. Our research group presented experimental contributions in the academic field [16,17]. However, the quantification of the loss of benefits for such systems due to the recent introduction of very thick insulation layers interposed between the inner mass and the cooling air gap is still lacking.

The present subsection begins with experimental investigations on two different ventilated walls to deepen the influence of some of the identified features on the walls performance through measures on real buildings. It ends with simulations to extend the study to other wall configurations.

Table 5.2 Two ventilated walls experimentally investigated

	Zinc-titanium wall Conservative	Clay wall Selective
Thickness (cm)	44.5	36
External cladding	Titanium + wooden plank	Clay
Thickness (cm)	2.5	4
Air gap	6	6
External mass	Semisolid brick	—
Thickness (cm)	12	—
Insulation	XPS	EPS
Thickness (cm)	7	4
Internal mass	Hollow brick	Clay block
Thickness (cm)	12	20
Internal finish.	Plaster	Plaster
Thickness (cm)	1.5	1.5
U (W/(m^2 K))	0.28 (0.31)[a]	0.39 (0.42)[a]
Y_{12} (W/(m^2 K))	0.04 (0.085)	0.051 (0.061)
κ_1 (kJ/(m^2 K))	41 (41)	39 (39)
f (−)	0.22 (0.27)	0.14 (0.27)
Δt (h)	11.5 (10.8)	13.2 (10.8)
External cladding		
ε (−)	0.1	0.9
α (−)	0.3	0.7
λ	0.15 (wooden plank)	0.35 (hollow bricks)

[a]The values are calculated either considering an unventilated air layer or a well ventilated air layer (values in brackets), according to EN ISO 6946:2007. In the latter case the values are obtained by disregarding the layers between the air gap and the external environment, and including an external surface resistance corresponding to still air.

Consider as an example the two ventilated walls shown in Table 5.2. They are adopted in the outer envelope of buildings D2 and D1 in Appendix A. They present different external opaque cladding, realized, respectively, with a continuous airtight zinc-titanium and an open joints clay facing. Moreover, they are characterized by different positions of the massive layers respect to the external environment and an unequal position of the insulating layer within the stratigraphy. In the zinc–titanium wall, the ventilated cavity is confined toward the outside by a cladding with low inertia and on the inner side by a massive layer (thermal insulation is then laid behind it). In the clayed wall, the ventilated cavity is confined toward the outside by a massive layer and on the inner side by a low inertia insulation. So the two walls can be classified, respectively, as a "conservative" solution (zinc-titanium wall), in which the heat is stored within the massive layer and trapped thanks to an external low conductive layer and a heat dispersing solution (clay wall), more influenced by the external variations since the massive layer faces outside and for the air infiltration through the open joints. These behaviors remind the two categories drawn by R. Bahnam, namely, "conservative" and "selective" solutions [18].

Although it not possible to compare the two walls directly since they have a different thermal resistance for the unequal insulating layer thickness and for the different materials adopted, it is still possible to observe how the behavior of the two walls is influenced by the different massive-insulating layers relative position. The incidence of the chimney height and of the external microclimatic conditions were experimentally investigated for both walls and simultaneous comparisons were done for extensive periods. The plan for the instrument positioning is reported in Appendixes D.1 and D.2. Thermal simulations made it possible to compare the behavior of the two ventilated facades by equaling the stationary transmittances and extend the considerations to other ventilated facades types.

5.2.1.1 The stack effect

The natural movement of the air into the chimney is called stack effect and it is driven by buoyancy. Buoyancy is an upward force that occurs due to a difference in indoor-to-outdoor air density resulting mainly from temperature differences. The rising of air in the chimney creates a draught that draws air from below. The air velocity is expressed through the following equation [4] based on mass and energy balance:

$$v = \sqrt{\frac{2H(\rho_e - \rho_i)g}{\rho_i(f_r\frac{H}{d} + 1.5)}} \tag{5.1}$$

where v = mean air velocity in the cavity (m/s); H = height of the cavity (m); ρ_e, ρ_i = external and internal air density (kg/m^3) (depending on air temperature, pressure, and humidity); g = gravitational acceleration (m/s^2); f_r = coefficient of friction; and d = equivalent diameter (m) (chimney width).

The *air velocity in the ventilation channel* in the two walls will be strongly influenced by the chimney configuration (height, width, roughness of the materials adjacent to the gap) and by the difference in outside-to-inside air densities. Air density strictly depends on temperature, pressure and, only secondarily, on humidity. Hence, having fixed materials and geometries, the airflow (using the ideal gas law) mainly depends on the temperature difference ΔT between the air in the gap and the outdoor air: the greater the thermal difference in inside-to-outdoor air temperature ΔT, the greater the buoyancy force, and thus the stack effect. Pressure difference could play an important role too, especially under high windy regimes.

To understand the behavior of a wall (with fixed geometries and materials) it is important to analyze (also at the varying of wind regimes) the daily variation of the difference in temperature between the outside air and the air in the channel that is the driving force for the stack effect. The combined effects of the outdoor temperature plus the incident solar radiation on the external cladding influence the air temperature inside the channel. These effects are easily detectable through the plot of the sol-air temperature recorded in the external cladding.

5.2.1.2 The sol-air temperature

When a wall is struck by a solar radiation of intensity I (W/m^2), it absorbs part of the incident energy according to its solar absorbance coefficient α, producing a rise in temperature of its own surface. This temperature increase produces a variation of the heat flow exchanged with the external air respect to the situation of absence of solar radiation.

This situation could be treated as a not irradiated condition in which a layer of air adjacent to the material had a fictitious—more elevated—temperature, such as to provide the same heat flows obtained in the irradiated case.

The sol-air temperature (Fig. 5.2) expresses this fictitious parameter.

It is defined as the outside air temperature that, in the absence of solar radiation, would give the same temperature distribution and rate of heat transfer through the wall, as exists due to the combined effects of the actual outdoor temperature distribution plus the incident solar radiation. This parameter is calculated with the following relation [19]:

$$T_{\text{sol-air}} = T_a + \alpha \, I/h_e - \varepsilon \Delta R/h_e \tag{5.2}$$

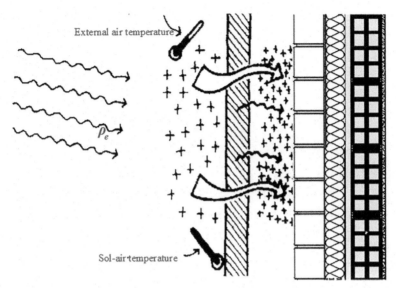

Figure 5.2 Sol-air temperature. *From Design Handbook on Passive Solar Heating and Natural Cooling (Habitat, 1990). Freely available on-line.*

where $T_{\text{sol-air}}$ = sol-air temperature (K); T_a = external air temperature (K); α = solar absorptance (–), values in Table 5.2; I = solar irradiance incident on the vertical surface (W/m²); and h_e = external surface heat transfer coefficient (W/(m² K)); ε = emissivity for infrared radiation (–), the values used are reported on Table 5.2; ΔR = difference between long wave radiation incident on surface from sky and surroundings and radiation emitted by black body at outdoor air temperature (W/m²). So the sol-air temperature is equal to the sum of air temperature plus the effect of incident radiation absorbed by the surface minus the effect of emitted radiation to the sky and surroundings. This latter term ($\varepsilon \, \Delta R/h_e$) is a long wave radiation factor that could be neglected for vertical surfaces.

5.2.2 Behavior of the zinc-titanium wall, experimental measures

The present section reports the experimental measures on a building (case study D.2 in Appendix A) that has ventilated facades with a zinc–titanium cladding (Table 5.2 and Figs. 5.3 and 5.4).

Figure 5.3 A school building in Ancona with zinc-titanium ventilated walls.

Figure 5.4 Detail of the zinc-titanium external cladding.

5.2.2.1 Behavior of the wall in summer, cloudy and sunny days

To verify the impact of high solar radiation combined with high air temperatures on the efficiency of the wall, two summer days were chosen characterized by either low or high sol-air temperature on the external cladding.

In the graph of Fig. 5.5 the sol-air temperature, the external surface temperature, the air temperature and velocity inside the ventilation channel, at two chimney heights (inlet at 1 m and top wall at 10 m) are reported. The external air temperature value and wind velocity are also reported.

The day 10th of July was characterized by cloudy and windy conditions (high wind from north in the central hours) and by low temperatures, while 08th of July was sunny with elevated temperatures and the wind velocities increased only in late afternoon.

On the cloudy day, *the external air temperature* ranged between 22°C and 27°C, while on the sunny day between 23°C and 35°C.

The *sol-air temperature* of the wall recorded at 10 m in the central hours of the day, ranged between 35°C and 45°C on the cloudy day with a variable trend, while between 45°C and 55°C on the sunny day.

The *external surface temperature* of the wall at 10 m was between 22°C and 47°C with a variable trend on the cloudy day and between 22°C and 55°C on the sunny day, with a difference of 8°C in the central hours of the day. The external surface temperature values recorded at a height of 1 m are more elevated for the effect of the ground-reflected radiation and since the cooling effect due to wind has major incidence on the higher portions of the wall.

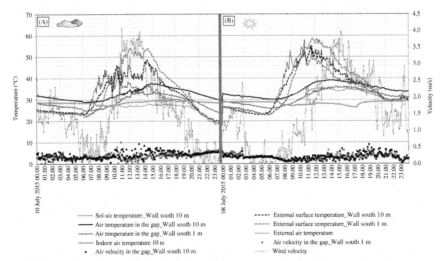

Figure 5.5 Summer experimentation on the zinc-titanium wall, comparison between (A) cloudy and (B) sunny days.

However, these different boundary conditions do not affect the inside air gap temperatures in a noticeable manner, since the values of air temperature in the channel at the top of the wall always remain higher than those recorded at 1 m of height.

The *air temperature in the ventilation channel* at an height of 10 m ranges between 29°C and 37°C on the cloudy day and between 30°C and 39°C on the sunny day, with a difference of about 2°C in the middle of the day. At an height of 1 m it ranges between about 25°C and 35°C in cloudy day and between 26°C and 36°C in the sunny one (about 1°C higher).

The *air velocity in the ventilation channel* at 10 m has slightly different trends in the two days depending on the difference in internal-to-external air temperature. In the selected cloudy day, this temperature difference ΔT is higher (up to 10°C in the central hours and 5°C during the night), while in the sunny day it reaches lower values (3–4°C in the central hours and about 5°C during the night). Therefore the average air velocity in the ventilation channel is slightly higher (up to 0.6 m/s) on the cloudy–windy day respect to the sunny day (0.2 m/s on average).

The different wall efficiency under unequal sun conditions is clearly visible in the central hours and in the late afternoon. In the middle of the cloudy day, the presence of wind from the north direction (backside of the wall) lowers the sol-air temperatures and determines a low-pressure area at 10 m. Under these conditions, the air temperature within the chimney assumes homogeneous values at the different heights (10 and 1 m) enhancing the buoyancy driven flow. In late afternoon of the cloudy day, the higher efficiency of the buoyancy force could be ascribed to higher temperature difference between outdoor air and air enclosed in the gap.

The presence of a massive brick layer just adjacent to the heated air gap determines that the increasing of the temperature within the gap has a flatter trend respect to that recorded for the external surface values. In both days, it is possible to note from the first hours of the morning (about 7.00 a.m.) up to about midafternoon an increase of the values of the air temperature in the gap. The external air temperature increases significantly in the first hour (from 7.00 a.m. to 8.00 a.m.), almost equaling the temperature inside the top part of the channel only in the sunny days. In such days, the consequent vanishing of the difference in internal-to-external air temperature causes a sudden reduction of the airflow rate, with velocity values being close to zero at the intersection of the curves.

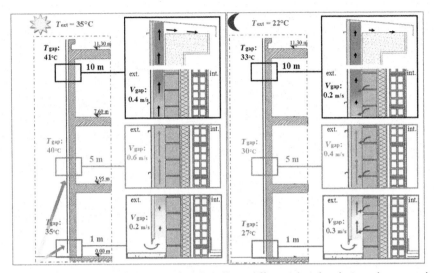

Figure 5.6 Behavior of the ventilated wall at different height during the central hours of the day and in the middle of the night. The schemes are derived by experimental measures on June 5.

In both sunny and cloudy days the outside temperature increases much less in the central hours (from 11.00 a.m. onward), resulting in an increasing divergence of the air temperature curves ($T_{outdoor} - T_{channel}$). This difference activates the stack effect and determines the consequent sudden increase of the values of air velocity. After 7.00 p.m. and during the nighttime the difference between the temperature in the channel and outside air temperature remains almost constant (about 5°C for both days).

The schemes in Fig. 5.6 highlight that the wall has an almost uniform behavior at the different heights, in both day and night.

5.2.2.2 Effect of the wind

A comparison between the air velocities in two days with very similar external temperatures and characterized by wind coming predominantly from the southern side but with different intensities, indicates the same trend for the wall (Fig. 5.7). Hence, the wind regimes do not influence the wall performance when coming frontally respect to the wall, since the facade has closed joints.

Using the values of air velocity in the ventilation channel in the two selected days, it was possible to determine the airflow conditions (Fig. 5.8). A preliminary analysis indicates a range in which no airflow

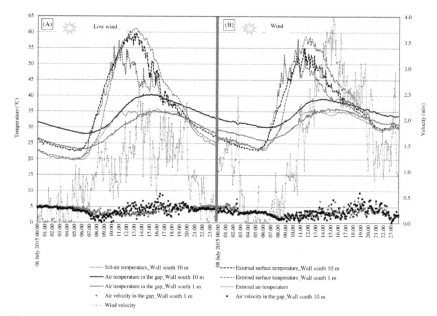

Figure 5.7 Summer experimentation on the zinc-titanium wall, comparison between (A) not windy and (B) windy days.

values are present. This gap represents a change in the airflow in the ventilation channel from laminar to turbulent. In both days, the wall has an almost homogeneous behavior with similar values at the two heights (1, 10 m) and mainly laminar flow (except for few central hours).

The study of two days with high wind regimes but different wind directions, clarifies that the wind has an incidence on the wall behavior only when coming from the northern side (comprised within a range of angles of 90 and 270 degrees, and thus falling between *dashed lines* in Fig. 5.9). High wind intensity from the back of the wall (northern building side) creates a depression in the top part of the wall (Fig. 5.10) thus enhancing the stack effect. This is visible in the increase of air velocity in the central hours of 10th of July.

5.2.2.3 Effect of the chimney height

A comparison of the ventilation regimes within the gap for different chimney heights (12 and 4 m) also at the varying of wind conditions is shown in Fig. 5.11. The measures were recorded at the central height of the two walls.

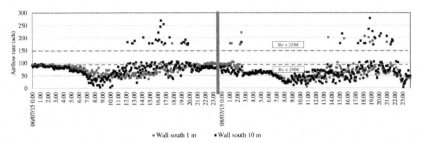

Figure 5.8 Summer experimentation on the zinc-titanium wall, airflow rates comparison between not windy and windy days.

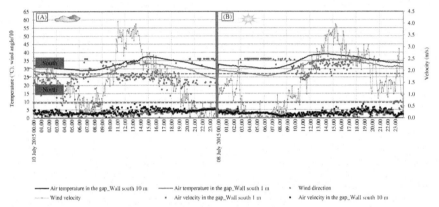

Figure 5.9 Summer experimentation on the zinc-titanium wall. Comparison of air temperature and velocity in the channel for different wind directions, (A) from the northern side on July 10 and (B) from the southern side on July 8.

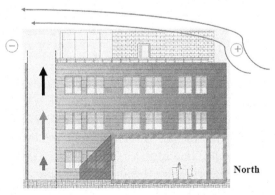

Figure 5.10 Schematic front view with enlarged wall section to highlight the effect of wind coming from the backside of the wall.

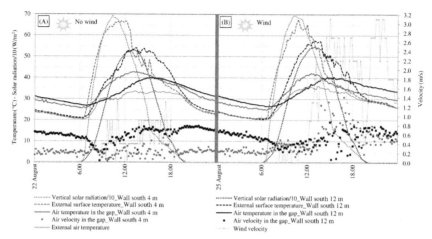

Figure 5.11 Summer experimentation on the zinc-titanium wall, comparison between two different chimney heights, 4 and 12 m, (A) in a not windy day and (B) in a windy day.

On both days the 4-m wall is subject to a higher sol-air temperature peaks than the 12-m wall. This difference is because the sensors positioned at the walls mid-height are at about 6 m above ground level in the higher one and at about 2 m above ground level in the lower one. Therefore the latter is affected by a stronger ground-reflected radiation. This different boundary condition affects the external surface temperature trends: the maximum temperature values measured on the surface of the 4-m wall are higher than those recorded for the 12-m wall.

The trend of air temperatures in the ventilation channel for the two walls is influenced by the different surface temperatures: in both the two days the lower wall presents higher values in the central hours of the day (up to 43°C), while the higher wall has a flatter trend (maximum value of 40°C). However, the 12-m wall has higher air velocities for more effective buoyancy activation.

The presence of wind from south has a small effect on the 12-m wall determining only a more scattered trend for the air velocity in the gap, thus confirming what found above, while very different airflows in the channel are recorded only for the lower wall. The minor influence of the wind on the higher wall is due to the increase of fluid dynamic resistance at the increasing of the chimney height. This result confirms through experimental data what has been obtained in an analytical way in the literature [4].

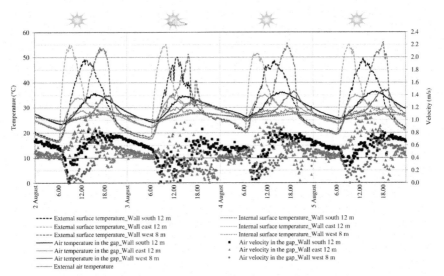

Figure 5.12 Summer experimentation on the zinc-titanium wall, comparison between three different exposures. *From F. Stazi, A. Vegliò, C. Di Perna, Experimental assessment of a zinc-titanium ventilated façade in a Mediterranean climate, Energy Build. 69 (February 2014) 525–534, ISSN 0378-7788.*

5.2.2.4 Effect of the wall exposure

The exposure of the wall causes only a slight shift in the onset of the stack effect.

To verify the effect of exposure to sunlight, three walls, south-, east- and west-facing, of, respectively, 12, 12, and 8 m, were compared (Fig. 5.12) for a period including both sunny and cloudy days (from the 2nd of August to the 5th of August). The external air temperature, the external surface temperature, the internal surface temperature, the air velocity and temperature in the ventilation channel are reported.

The selected days present similar external air temperature trends ranging between 18–20°C at night and reaching 28–30°C in the central hours of the day.

By comparing the external surface temperatures, it can be noted that the minimum values are recorded in all cases at around 6.00 a.m., while the maximum values are found at different times. On the east-facing wall, they are reached at 9.00 a.m., on the south-facing wall at 1.00 p.m., and on the west-facing wall at 5.00 p.m., reaching maximum values of 55°C, 50°C, and 56°C, respectively. Therefore even if the south-facing walls receive solar radiation for a longer time, the east- and west-facing walls have higher temperatures because of the low solar radiation angle in the first and last part of the day.

A comparison of the internal surface temperatures reveals that in all three cases the values are between 25°C and 28°C with only a slight time lag and similar trends.

The study of the air temperature values recorded inside the channel for the three walls and the difference ΔT between these values and the external air temperature (that is the main cause of stack effect activation) explains the different behavior of the air velocities on the three walls.

The east-facing wall is affected by direct solar radiation sooner than the other walls thus undergoes an immediate rise in temperature in the gap suddenly activating the stack effect with a maximum velocity value reached at about 12.00 a.m. Subsequently the temperature in the gap quickly decreases almost equaling the outdoor air temperature value (that is still high) at 6.00 p.m., thus determining a considerable reduction in the stack effect with a consequent scarce performance during the night if compared to the other two walls.

In the south-facing wall, the air temperature in the channel increases more gradually, and in the early morning hours it follows the trend of the outside temperature therefore with a resulting sudden decrease of the buoyancy effect. The maximum air velocity in the channel in this case is at about 4.00 p.m. with a subsequent gradual reduction. This wall preserves higher air temperature values in the channel than the other two walls throughout the whole day, allowing a better activation of the stack effect.

In the west-facing wall, the stack effect is activated later than the other two walls and since it has a lower chimney height, it reaches lower maximum air velocity in the channel.

In conclusion, the exposure affects the external surface temperature trends, modifying the peak time and the amplitude of the plot. Therefore have a different impact on the values of velocity in the channel: higher for the east-facing wall in the morning and higher for the west- and south-facing walls during the rest of the day, with a more beneficial nighttime cooling effect for the latter.

5.2.2.5 Thermal fluxes with and without cooling system

Selecting two days with (Monday) and without (Sunday) the use of the cooling system it is possible to study the different heat fluxes (Fig. 5.13).

On Sunday, the trend of the heat flux initially follows the trend of air velocity in the gap. The decrease of the airflow in the early morning

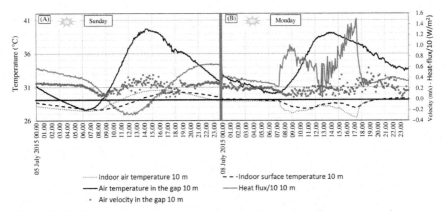

Figure 5.13 Winter experimentation on the zinc-titanium wall, comparison between two days with cooling system (A) switched off (Sunday July 5) and (B) switched on (Monday July 8).

and its sudden nullification at 8 a.m. corresponds to the reduction of the entering heat flux through the wall, even up to the changing of its direction, exactly at 9 a.m. After this hour, the flux is outward demonstrating the cooling effect that lasts up to 3 p.m. and determines a reduction on the surface temperature respect to the indoor air temperature. Then the decrease of the inside air temperature during the evening determines a reduction of the outgoing flux and its change of direction.

On Monday, the system is turned on from 7 a.m. to 5 p.m.: in that period the indoor temperature is reduced by 3°C (respect to the unconditioned situation) and the flux is always directed from the wall to the indoor, also influencing the surface temperature trend. The air temperature in the gap is 1°C lower than the value recorded in not cooled conditions.

5.2.2.6 Behavior of the wall in winter days

In winter, the same observation drawn for the summer period could be adopted. The main driving force for the stack effect is the temperature difference between air in the gap and outdoor. This temperature gradient in two days with different incident direct radiation (Fig. 5.14) is similar except in the very central hours. In such hours in the cloudy day, the increase of wind velocities (mostly from north) determines an effective activation of the buoyancy force. In unheated conditions (on December

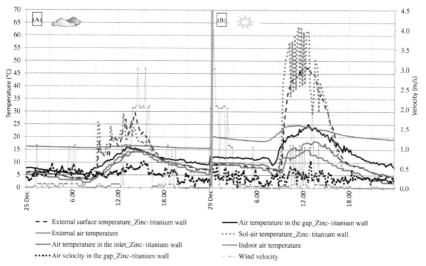

Figure 5.14 Winter experimentation on the zinc-titanium wall, comparison between (A) cloudy and (B) sunny days.

25), the internal air temperatures are stable around 16°C since the presence of an external "passively heated" gap reduces the outgoing fluxes. This behavior is beneficial also under heated conditions for the reduction of the heating system work.

5.2.3 Behavior of the clay-cladded wall, experimental measures

The present section reports the experimental measures on a building (case study D.1 in Appendix A) that has ventilated facades with a clay cladding (Table 5.2 and Figs. 5.15 and 5.16).

5.2.3.1 Behavior of the wall in summer, cloudy and sunny days

To verify the impact of direct solar radiation on the efficiency of the wall, two summer days were chosen with similar wind conditions (low wind speed) but characterized by either the presence or the absence of direct solar radiation.

The graph in Fig. 5.17 reports the sol-air temperature, the external surface temperature, the air temperature, and velocity inside the ventilation channel, at two chimney heights (inlet at 1 m and top wall at 10 m). It includes the external air temperature values and wind velocities, too.

Figure 5.15 The monitored building with the clay-cladded ventilated wall. Northern side.

Figure 5.16 The monitored building with the clay-cladded ventilated wall. Southern side.

The external air temperature ranged between 22°C and 32°C regardless the cloudiness.

The *sol-air temperature* of the wall recorded at 10 m in the central hours of the day (9 a.m.−3 p.m.) was very variable on the cloudy day, reaching high values only for 1−2 hours, while it was greater on the sunny day.

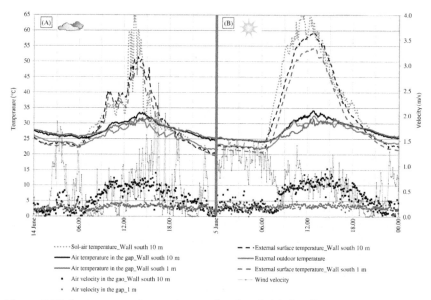

Figure 5.17 Summer experimentation on the clay-cladded wall, comparison between (A) cloudy and (B) sunny days, southern exposure.

The *external surface temperature* of the wall at 10 m reached 50°C with a variable trend on the cloudy day and 60°C on the sunny day, with a difference of 10°C in the middle of the day. The values of surface temperature at 1 m are similar to those recorded in the top part of the wall since no effect of the ground-reflected radiation occurred for the presence of a nonreflecting floor adjacent to the wall (gravel on the roof of the lower volume).

The *air temperature in the ventilation channel* at 10 m ranges between 25°C and 35°C in both days; at height of 1 m it ranges between about 26°C and 30°C in both days. Therefore the curves for the temperature at the two chimney levels intersect with each other two times, at 9 a.m. and at 6 p.m. In the morning before 9 a.m., the air in lower part of the chimney records slightly higher temperatures. After 9 a.m., the air in the upper part undergoes a sudden increase thus activating an effective ventilation (velocity up to 1 m/s) and remains higher until 6 p.m.; after this hour the upper air temperature suddenly decrease becoming lower than those recorded at 1 m of height and determining a strong reduction of the chimney effect. During the whole night, the air velocity remains at 0.2 m/s (Fig. 5.18).

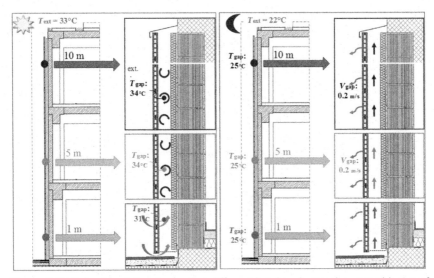

Figure 5.18 Behavior of the clay wall at different heights during the central hours of the day and in the night.

The wall efficiency under unequal sun conditions is very similar.

In an overall evaluation, in both cloudy and sunny days the *air temperature at the inlet openings* recorded, in the morning and in evening, values similar to that recorded at the top of the wall and constantly 2.5°C higher than the external temperature. In the central hours, the upper air overheats respect to the inferior air, thus activating the chimney effect. In such constructive technique, the strong influence of the external environment on the air gap is clearly visible. The massive cladding could not benefice from its storage ability being directly in contact with the outside. Moreover the presence of an insulating layer that subdivide the gap from the inner mass determines that the heat inside the gap is fully lost toward the outside at night, while the temperatures increase with the same rate of the external temperature during the day. No time shifting is highlighted.

5.2.3.2 Effect of the wind

The presence of wind frontally hitting the open joint southern wall (Fig. 5.19), instead of enhancing the chimney effect, determines a more scattering trend for the air velocity in the central hours of the day thus reducing the wall performance. When the wind velocities become higher

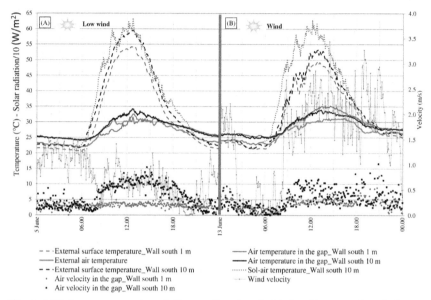

Figure 5.19 Summer experimentation on the clay-cladded wall, comparison between days with (A) low wind and (B) high wind from south.

than a value of about 2 m/s, the wall presents inverse ventilation clearly visible by the temperatures increase in the lower part that exceeds the values recorded at the top.

Using the values of air velocity in the ventilation channel in the two selected days, it was possible to determine the airflow conditions (Fig. 5.20). In the sunny day without wind, the wall has very different values at the two heights (1, 10 m): a mainly laminar flow is highlighted in the lower part, while in the top of the wall a turbulent flow occurred in the central hours of the day. In the windy–cloudy day, the air velocity presented a more variable trend for the airflow inversion.

5.2.3.3 Effect of the chimney height

The temperature for the various layers of two south-facing walls with a ventilation channel, respectively, of 6- and 12-m wall were compared at various times during the day to highlight the different performance depending on the chimney height (Fig. 5.21). The results show that during the central hours of the day, the temperature of the air in the gap of the 12-m wall is considerably higher than the temperature found for the 6-m wall, while during the nighttime it is lower.

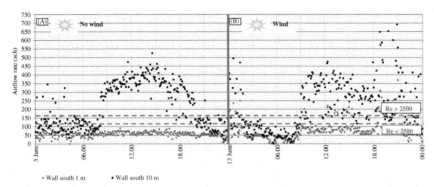

Figure 5.20 Summer experimentation on the clay-cladded wall, airflow comparison between days with (A) low wind and (B) high wind from south.

The maximum external surface temperature for both walls was found at midday with a difference of 5°C between the walls (39°C for the 6-m wall and 44°C for the 12-m wall). At 3.00 p.m., the baffle becomes cooler, due to the chimney effect, which is proportionally greater for the higher wall (the temperature goes down by 3−4°C for the 12-m wall between 12.00 a.m. and 3.00 p.m. compared with a reduction of about 1°C for the 6 m wall). This phenomenon is even more noticeable at 6.00 p.m. when the external surface temperature of the 12-m baffle is 30°C against 31°C of the 6-m wall.

The greatest cooling effect of the higher wall is demonstrated by the comparison of air velocities in the channels (Fig. 5.22). The value recorded for the 6-m wall during the hottest hours of the day is around 0.5−0.6 m/s, while it rises up to 0.8−0.9 m/s for the 12-m wall. The maximum velocity values are found from 11.00 a.m. to 3.00 p.m. for both the walls.

5.2.3.4 Effect of the exposure

To verify the impact of the height of the solar angle, the external and internal surface temperatures of the 6-m ventilated wall were compared for different exposure (east, south, west) with floating internal temperatures (air conditioning closed).

The graph concerning the external surface temperatures (Fig. 5.23) shows that the minimum values are recorded between 3.00 a.m. and 5.00 a.m. for all the walls. The maximum values are found on the wall facing east between 8.00 a.m. and 10.00 a.m., on the wall facing south between

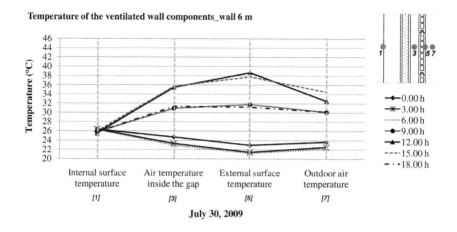

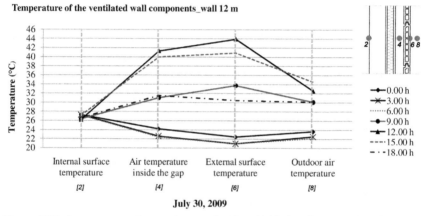

Figure 5.21 Summer experimentation on the clay-cladded wall, comparison between a 6-m height wall and a 12-m height wall. *From F. Stazi, F. Tomassoni, A. Vegliò, C. Di Perna, Experimental evaluation of ventilated walls with an external clay cladding, Renewable Energy 36 (12) (December 2011) 3373–3385, ISSN 0960-1481.*

11.00 a.m. and 1.00 p.m., and on the wall facing west between 3.00 p.m. and 5.00 p.m., reaching 46°C, 41°C, and 48°C, respectively. Therefore even if the walls to the south are exposed to solar radiation for a longer period of time, the walls facing east and west have higher temperatures. In particular, the west-facing wall reaches the greatest values since the solar radiation strikes the wall with a low angle in the later hours of the afternoon.

A comparison of the internal surface temperatures shows that the ventilated wall lowers the maximum temperatures by about 22°C on the west

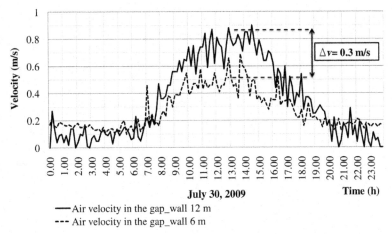

Air velocity in the gap (m/s)

Figure 5.22 Air velocity in the gap at the varying of chimney height. *From F. Stazi, F. Tomassoni, A. Vegliò, C. Di Perna, Experimental evaluation of ventilated walls with an external clay cladding, Renewable Energy 36 (12) (December 2011) 3373–3385, ISSN 0960-1481.*

face, 15°C on the south face, and about 19°C on the east face, bringing them within the 26–27°C range in all three cases. On the east and west sides the minimum values are found at 6.00 a.m., while the maximum temperatures, which are slightly higher on the east-facing wall, are recorded at around 2.00 p.m. The south-facing wall behaves very differently having a flatter trend: the minimum value is recorded at 10.00 a.m. and the maximum internal surface temperature is found at around 10.00 p.m.

5.2.3.5 Thermal fluxes with and without cooling system

The comparison between two days with the cooling system working (Wednesday) and switched off (Saturday) shows (Fig. 5.24) that on both days the surface temperature remains higher than the internal one (thus with entering fluxes) except for the central hours. The main effect of the cooling system is the reduction of the hours in which the flux is outgoing. As a matter of facts on Wednesday, with cooling system working, the internal air and surface temperatures are both reduced to about 1.5°C and the period in which the air temperature is higher than the surface temperature is shortened.

Comparison between east, south, west external surface temperature

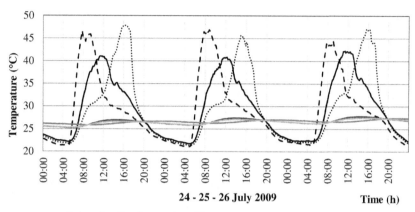

Figure 5.23 External and internal surface temperatures at the varying of wall exposure. *From F. Stazi, F. Tomassoni, A. Vegliò, C. Di Perna, Experimental evaluation of ventilated walls with an external clay cladding, Renewable Energy 36 (12) (December 2011) 3373–3385, ISSN 0960-1481.*

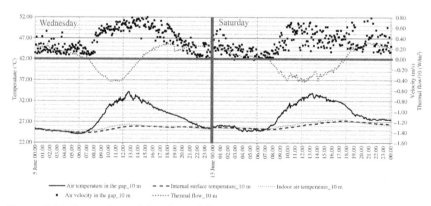

Figure 5.24 Comparison of the thermal fluxes with cooling system open and closed.

5.2.3.6 Behavior of the wall in winter days

In winter, the study of the behavior of the 6-m wall allows to adopt the same observations drawn for the summer period. The direct radiation has a negligible incidence on the wall efficiency (Fig. 5.25), while the high wind velocities interact with the chimney effect in the gap increasing the airflow. High wind regimes striking frontally the wall (as occurred in the early morning of the sunny day) determine an inverse airflow direction causing disordered airflow.

5.2.4 Comparison, simultaneous experimental measures

The study of the whole summer period for a south-facing zinc-titanium cladded wall ($h = 12$ m) (Figs. 5.26 and 5.27) clearly highlights that the average air velocity in the gap is around $0.2-0.3$ m/s with similar trends at the inlet and in the top part of the wall. Airflow reductions are recorded in the morning (between 6 a.m. and 12 a.m.) for the increasing

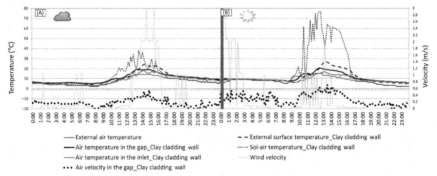

Figure 5.25 Winter experimentation on the clay-cladded wall, comparison between (A) cloudy and (B) sunny days.

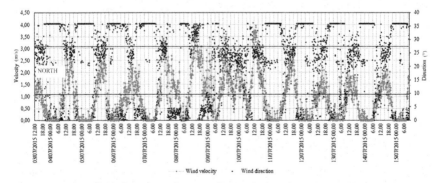

Figure 5.26 Weather data regarding wind recorded in the summer experimentation on zinc-titanium cladding.

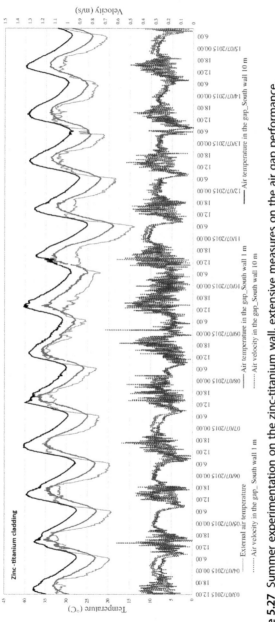

Figure 5.27 Summer experimentation on the zinc-titanium wall, extensive measures on the air gap performance.

of the external air not suddenly followed by the internal air. The slower increase of the temperature within the gap is due to the presence of a conservative air gap, characterized by a low conductive cladding and a storing massive layer enclosed behind it. The temperatures recorded in the lower and upper part of the gap have parallel trends demonstrating the presence of a homogeneous behavior of the gap due to the storing mass.

In the clay-cladded wall (Figs. 5.28 and 5.29) the air gap has a different insulation-mass configuration respect to the zinc wall, being (adopting the Banham classification) more "selective," admitting one aspect of the external environment (in this case the infiltration of air through the joints). The cladding is massive but outfacing. So during the night it follows the external temperature reduction with the same rate (at both chimney height, 1 and 10 m), while in the central day hours it shows a differential behavior at the two heights since the buoyancy flow activation cools the lower part of the chimney moving the hot air in the upper areas. In some days (June 12 and 13) high wind from south hitting the open joint frontally creates a disordered air flux.

The scheme in Fig. 5.30 highlights an almost constant behavior for the zinc-titanium wall throughout the day and night, while a more variable efficiency in the clay-cladded facade.

Fig. 5.31 reports the external air temperature, the sol-air temperature, the air temperature, and the velocity of the air in the channel for both the clay wall and zinc-titanium wall, both on the southern building side. To compare the walls similar days were selected (obtained from a June monitoring for the clay wall and from a July monitoring for the zinc wall) characterized by either low or high wind velocities. The windy days were selected by choosing the most impacting wind directions for each wall, frontally (south) for clayed open joint wall and from the backside (north) in the case of the wall with airtight metal cladding.

The difference between *the external surface temperatures* in zinc-titanium and clay cladding at 10 m is particularly noticeable during the middle of a typical (low windy) day when the clay cladding undergoes to a considerable thermal overheating (up to 60°C), for its higher absorbed solar energy (since characterized by a higher absorptivity α for the solar radiation). However, in the windy day the curves are equaled since at 10 m height the cooling effect of wind has more influence on the clay cladding that has a very higher emissivity than the metal surface, indicating a higher rate at which the heat leaves the cladding. As demonstrated in

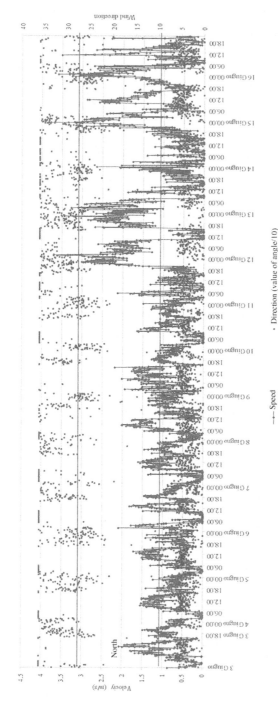

Figure 5.28 Weather data regarding wind recorded in the summer experimentation on clay cladding.

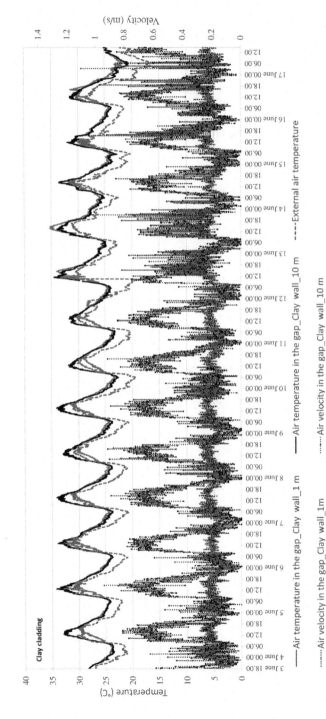

Figure 5.29 Summer experimentation on the clay-cladded wall, extensive measures on the air gap performance.

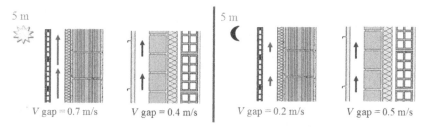

Figure 5.30 Comparison of the two facades at middle height in the central hours of the day and in the night.

Section 2.5.3 under high ventilation rate the effect of the thermal mass is nullified, whatever its position.

The conservative behavior of the zinc cladding is the cause of the higher temperatures recorded inside its channel and of the time shift between the maximum of the external surface temperature and the maximum of the internal air gap temperature since the enclosed mass contributes with its storage ability.

In the zinc wall the difference between the air temperature in the gap and outside air temperature is high throughout the day (approx. 10°C), activating the chimney effect almost constantly but with lower efficacy than the clayed wall for the higher roughness of its internal surfaces in the gap and for the air tightness of the cladding which excludes the air infiltration. The clay cladding strongly lowers its efficacy after 6.00 p.m. and for the whole night phase since the mass of the external cladding does not retain heat, thus strongly reducing the stack effect throughout this period of time.

In conclusion the different relative positions between the insulation and mass in the two walls result in a more efficient stack effect activation during the late afternoon and during the night for the zinc-titanium wall, while in the central hours of the day for the clay-cladded wall.

In winter, a simultaneous monitoring of the walls shows that the airflow is almost constant in the zinc-titanium wall (Figs. 5.32 and 5.33) with velocities around 0.25 m/s, while it is variable in the clay-cladded wall ranging between 0 and 0.5 m/s. In the days with high wind velocities from southern (or south-west) side (from 25/12 to 28/12, 1/01 and 08/1), the air temperature in the clay walls exceeds that recorded in the titanium one. This is because this wind direction has impact only in the former wall increasing its air velocities up to 0.8−1 m/s within the gap.

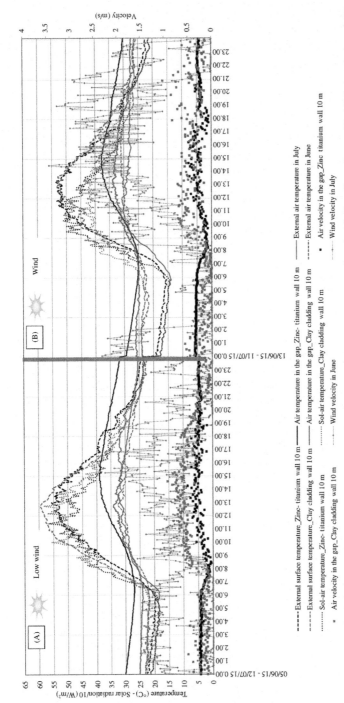

Figure 5.31 Summer experimental comparison between the zinc-titanium wall and clay-cladded wall in days with (A) low wind and (B) high wind.

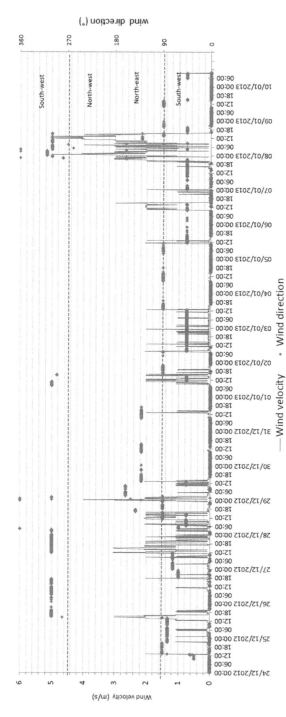

Figure 5.32 Weather data regarding wind recorded in the winter experimentation.

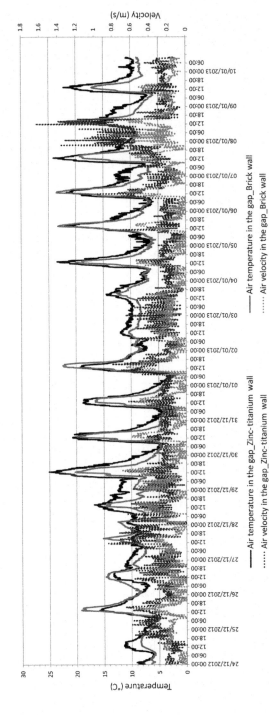

Figure 5.33 Extensive comparison between zinc-titanium wall and clay-cladded wall in winter.

5.2.5 Environmental sustainability

The study of the environmental impacts regarding preuse and postuse phase, by assuming as functional unit a square meter of wall and by adopting different methods (Fig. 5.34) demonstrates that the zinc-titanium wall has the greatest burdens and that the biggest weights are the construction materials according to all methods.

The detailed study of the burdens focusing on the construction phase (Fig. 5.35, Tables 5.3 and 5.4) demonstrates that for the zinc-titanium wall, the metallic cladding and the clay bricks of the inner layers (both perforated and semisolid) have the major impacts, with the highest burdens on human health (Carcinogens) and on resources (Fossil fuels). These environmental problems are due, for both materials, to the release of carcinogenic chemicals harmful for toxicological risk, and in resources category for the fossil fuel required to the production processes (in the former for the cold forging of metal sheets, in the latter for the fire brick kilns). The insulation is another impacting material (due to the industrial sintering process), but with smaller incidence.

For the clay-cladded wall, the highest total damage is caused by the clay blocks mainly weighting on Resource category. The main problems can be ascribed to Fossil fuel large consumption for brick kilns, as clearly visible in Table 5.4.

5.2.6 Optimal solution

As demonstrated above the materials of the layers adjacent to the ventilation channel and their position respect to external environment strictly influence the behavior and the efficacy of the vented facades. Several configurations are present in the market by changing the type of materials lying adjacent to the ventilation channel, thus obtaining very different winter and summer behavior. This section deepens this topic by varying through simulation (on model calibrated with measures) the walls configuration and equaling the thermal resistance.

Consider the ventilated walls shown in Table 5.5 adopting a gap width of 6 cm and the materials detailed in Table 5.6. The wall named Tit_1 is that experimentally analyzed in Section 5.2.2. The walls named Tit_2_3_4 concern the changing of the materials for the internal layers leaving the external cladding unchanged. In addition, two traditional walls, namely, Trad_1 and Trad_2 are reported for the sake of

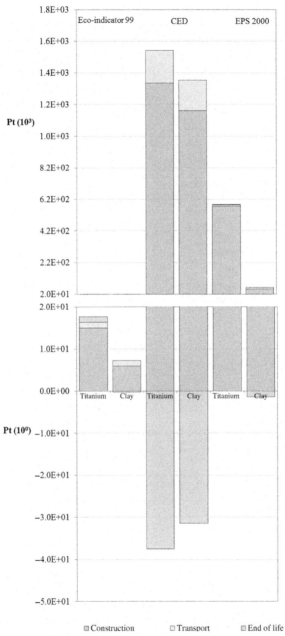

Figure 5.34 Life cycle phases results of two walls during 75-year life span with Eco-indicator 99, CED, and EPS 2000 methods.

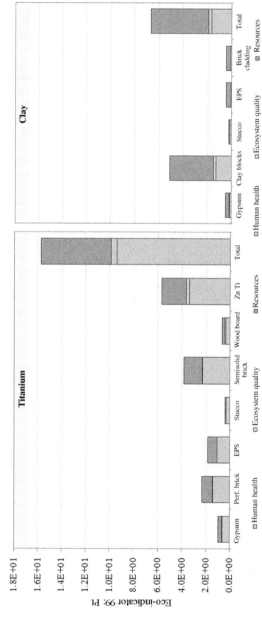

Figure 5.35 Construction phase results of two walls at year 0 with Eco-indicator 99.

Table 5.3 Detailed results of the life cycle impact assessment of the zinc-titanium wall in the construction phase

Zinc-titanium Eco-indicator 99 (H)—hierarchist perspective. Impact (Pt)		Gypsum	Perf. bricks	EPS	Stucco	Semisolid bricks	Wood board	Zn Ti	Total
Human health	Carcinogens	4.82E − 01	1.18E + 00	9.31E − 01	2.02E − 01	1.91E + 00	3.26E − 01	2.84E + 00	7.87E + 00
	Respiratory organics	3.57E − 03	6.45E − 03	5.76E − 04	6.16E − 04	1.05E − 02	4.11E − 03	2.56E − 02	5.14E − 02
	Respiratory inorganics	9.13E − 05	3.61E − 04	1.54E − 04	1.87E − 05	5.88E − 04	1.85E − 04	2.36E − 04	1.63E − 03
	Climate change	1.06E − 01	1.41E − 01	7.93E − 02	9.26E − 02	2.30E − 01	1.40E − 01	4.27E − 01	1.22E + 00
	Radiation	2.21E − 02	7.66E − 02	2.26E − 02	5.39E − 03	1.25E − 01	−1.09E − 01	9.41E − 02	2.37E − 01
	Ozone layer	1.22E − 05	2.79E − 05	4.31E − 09	3.68E − 06	4.53E − 05	6.55E − 05	2.00E − 05	1.75E − 04
Ecosystem quality	Ecotoxicity	1.23E − 02	7.16E − 03	2.82E − 04	1.34E − 03	1.17E − 02	1.39E − 02	1.36E − 01	1.82E − 01
	Acidific./ eutroph.	7.42E − 03	2.04E − 02	1.16E − 02	1.23E − 03	3.31E − 02	1.16E − 02	5.67E − 02	1.42E − 01
	Land use	1.26E − 02	1.94E − 02	6.92E − 06	−1.32E − 04	3.15E − 02	3.82E − 02	6.21E − 02	1.64E − 01
Resources	Minerals	2.22E − 03	8.70E − 03	4.48E − 05	3.96E − 04	1.42E − 02	1.70E − 03	1.13E + 00	1.16E + 00
	Fossil fuels	3.15E − 01	8.94E − 01	8.17E − 01	1.00E − 01	1.46E + 00	2.22E − 01	9.08E − 01	4.71E + 00
Total impact (Pt)		9.64E − 01	2.35E + 00	1.86E + 00	4.03E − 01	3.82E + 00	6.50E − 01	5.68E + 00	1.57E + 01

Conversion factors. DALY TO Pt: $\star$65.1 DALY$^{-1}\star$300Pt; PDFm^2/year TO Pt: $\star$1.95E − 4 Pm^2/year$\star$400Pt; $\star$1.19E − 4 MJ S.$^{-1}\star$300Pt; MJ Surplus TO Pt: $\star$1.19E − 4 MJ S.$^{-1}\star$300Pt.

Table 5.4 Detailed results of the life cycle impact assessment of the clay wall in the construction phase

Clay Wall Eco-indicator 99 (H)—hierarchist perspective. Impact (Pt)		Gypsum	Clay blocks	Stucco	EPS	Brick cladding	Total
Human health	Carcinogens	3.57E − 03	3.19E − 02	6.84E − 04	5.61E − 04	2.40E − 03	3.91E − 02
	Respiratory organics	9.13E − 05	9.90E − 04	2.07E − 05	1.24E − 04	1.34E − 04	1.36E − 03
	Respiratory inorganics	1.06E − 01	9.33E − 01	1.03E − 01	4.28E − 02	5.26E − 02	1.24E + 00
	Climate change	2.21E − 02	3.03E − 01	5.98E − 03	1.11E − 02	2.85E − 02	3.71E − 01
	Radiation	4.35E − 04	2.49E − 03	1.49E − 04	1.48E − 07	2.85E − 04	3.36E − 03
	Ozone layer	1.22E − 05	1.32E − 04	4.09E − 06	4.24E − 09	1.04E − 05	1.59E − 04
Ecosystem quality	Ecotoxicity	1.23E − 02	4.26E − 02	1.49E − 03	1.88E − 04	2.66E − 03	5.93E − 02
	Acidific./eutroph.	7.42E − 03	1.06E − 01	1.37E − 03	6.05E − 03	7.57E − 03	1.28E − 01
	Land use	1.26E − 02	7.75E − 02	− 1.47E − 04	6.72E − 06	7.20E − 03	9.72E − 02
Resources	Minerals	2.22E − 03	5.44E − 02	4.40E − 04	2.71E − 05	3.24E − 03	6.03E − 02
	Fossil fuels	3.15E − 01	3.57E + 00	1.11E − 01	3.88E − 01	3.33E − 01	4.72E + 00
Total impact (Pt)		4.82E − 01	5.12E + 00	2.24E − 01	4.49E − 01	4.37E − 01	6.71E + 00

Conversion factors. DALY TO Pt: $\star$65.1 DALY^{-1}$\star$300Pt; PDFm^{2}yr TO Pt: $\star$1.95E-4 Pm^{2}yr^{-1}$\star$400Pt; $\star$1.19E-4 MJ S.$^{-1}$$\star$300Pt; MJ Surplus TO Pt: $\star$1.19E-4 MJ S.$^{-1}$$\star$300.

Table 5.5 Analytical variations on ventilated wall types and relative dynamic parameters

	Tit_1 (as built)	Tit_2	Tit_3	Tit_4	Trad_1	Trad_2
Thickness (cm)	44.5	35.2	40.2	39.2	36	37
Ext cladding	Titanium	Titanium	Titanium	Titanium	–	Plaster
Thickness (cm)	2.5	2.5	2.5	2.5	–	1.5
External mass	Semisolid brick	–	–	–	Semisolid brick	–
Thickness (cm)	12	–	–	–	12	–
Insulation	XPS	XPS	XPS	XPS	XPS	XPS
Thickness (cm)	7	9	7	7	7	7
Internal mass	Hollow brick	Hollow brick	Hollow block	Hollow block	Hollow brick	Clay block
Thickness (cm)	12	12	20	20	12	20
Internal finish.	Plaster	Plaster	Plaster	Plaster	Plaster	Plaster
U^a (W/(m^2 K))	0.31	0.36	0.30	0.29 (0.71)[b]	0.31	0.30
Y_{12} (W/(m^2 K))	0.085	0.205	0.034	0.036 (0.24)[b]	0.085	0.034
κ_1 (kJ/m^2 K)	41	42	38	38 (41)[b]	41	38
f (−)	0.27	0.56	0.11	0.1 (0.30)[b]	0.27	0.11
Δt (h)	10.8	5.4	12.9	13.6 (9.30)[b]	10.8	12.9

[a]Calculated according to UNI EN ISO 6946:2007, by disregarding the thermal resistance of the air layer and all other layers between the air layer and the external environment, and including an external surface resistance corresponding to still air.

[b]The values in brackets regard the summer season when the ventilation in the gap is activated through the opening of the vents.

Table 5.6 Detail of the materials adopted for the walls simulated

Material	λ (W/(m K))	ρ (kg/m³)	c (J/(kg K))
Titanium cladding[a]	109	7100	400
Clay cladding[b]	0.5	792	814
Semisolid brick	0.28	775	1000
XPS insulation	0.039	30	1400
EPS insulation	0.04	35	1400
Hollow brick	0.24	535	1000
Fired clay block	0.18	840	1000
Plaster	0.4	1000	1000

[a]The titanium cladding is structurally embraced by an adjacent wooden batten 2.4 cm thick. Other interesting parameters are: solar radiation absorptivity = 0.3 (−), emissivity ε = 0.1 (−).
[b]The clay cladding presents the following parameters: solar radiation absorptivity = 0.7 (−), emissivity ε = 0.9 (−).

comparison: the former regards a cavity wall with insulation inside the gap, the latter a wall with clay blocks and external insulation layer.

All the vented solutions except Tit_4 are designed to have an external air gap ventilated throughout the year. The wall Tit_4 is the external ventilated insulation presented in Section 1.2.2 (Figs. 1.5 and 1.6) and deeply analyzed as retrofit measure in Sections 3.4.1 (solutions named C1 vent, C2 vent, and C3 vent) and 3.6.3 (solutions named W2, W4, and W6).

The selected walls differ for the relative position between massive and insulating layers and for the type of internal massive layer adopted (clay blocks or perforated bricks). They have almost the same thermal transmittance that assumes low values (about 0.30 W/m² K) and present very similar intermediate κ_1 value (near 40 kJ/m² K), but have different configurations regarding the alternation of massive and insulating layers and as a consequence differ for the attenuating attitude (f value).

The dynamic simulations highlight that the beneficial contribution of the chimney effect is relevant. The comparison of the internal surface temperatures for the various types of zinc-cladded walls in summer in a temperate Mediterranean climate (Fig. 5.36) shows that all the vented solution curves stand lower than the traditional ones. Solution Tit_3 obtained by introducing an external layer in traditional wall with external insulation layer (Trad_2) lowers the summer surface temperatures of 3°C respect to its traditional counterpart. In winter, this vented solution is slightly worse than Trad_2 for the exclusion of useful solar heat gains.

Preliminary experimentations recently carried out by our research group reveal that the values of the thermal resistance are slightly

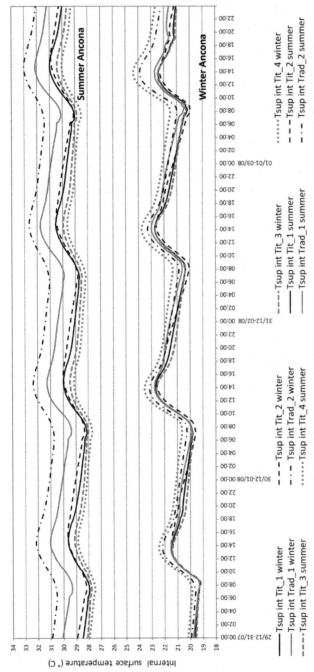

Figure 5.36 Internal surface temperatures for the various types of zinc-cladded walls in summer and winter seasons, Ancona.

underestimated by the standard EN ISO 6946, that for well-ventilated walls disregard the thermal resistance of the air gap and all other layers between the gap and external environment (including an external surface resistance corresponding to still air). In fact the on-site thermal transmittance of the as built wall Tit_1 (measured according to ISO 9869) results very lower than the calculated one (being around 0.18 W/m² K instead of 0.31 W/m² K). This observation is consistent with the winter experimental results (Section 5.2.2.6) that highlighted the crossing flow reduction for the presence of an external "passive heated" air gap.

The wall Tit_4 showed the best performance in the dynamical simulations. Indeed, while in all the other ventilated walls the air gap is external respect to the insulation layer, in Tit_4 case the air gap is inside and the chimney effect is activated only in summer. So in summer its great attitude to be crossed by the heat (high f value) is beneficial since the heat, coming mainly from inside, can be removed through the chimney effect. The ventilated wall Tit_4 has the best performance also in winter, thanks to its lowest both U value and decrement factor. However, more research is needed since no experimental data are available on this topic.

The extension to other extreme climates (Fig. 5.37) shows the same best and worst solutions and demonstrates that the choice of an appropriate solution presents almost the same benefices founded for the temperate climates, lowering the temperatures of about 3°C (from the worst to the best solution) in the very hot summer of Cairo. The vented solutions have instead equal or reduced performance than a traditional envelope in the very cold climate of Bolzano.

5.3 SOLAR WALLS AND TROMBE WALLS

5.3.1 Problem description

The recent increasing interest in passive solar systems is motivated by their capability of providing energy saving and thermal comfort. As a consequence their use is encouraged by European Directives [1,2] and national regulations of the Member States.

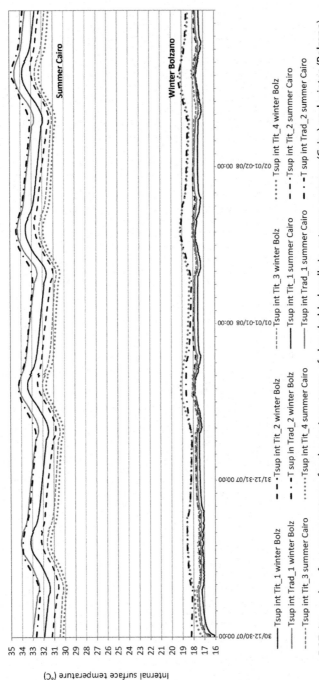

Figure 5.37 Internal surface temperatures for the various types of clay-cladded walls in extreme summer (Cairo) and winter (Bolzano).

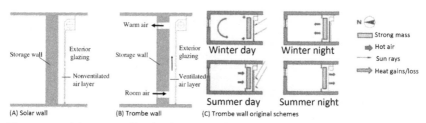

Figure 5.38 (A) Nonventilated solar wall; Classical Trombe wall with air thermocirculation (B); original schemes elaborated at the system conception (C).

Among the different *passive solar systems*, solar walls and Trombe walls have aroused the greatest attraction for their simple design and because even if they were born with the purpose of the exploitation of solar energy in cold climates and at high altitudes, they were successfully adopted even in hot climates since used as passive cooling system too. Both of them have an external glass and an internal mass with storage function. The external cavity could be shaded (e.g., by roller shutters) and, in the case of Trombe walls, could be also ventilated.

The introduction of a superinsulation layer on such walls born as winter sun capturing systems, would exclude the beneficial heat gains. Moreover, the adoption of thick insulations on the building components other than the southern wall (the roof, the ground floor slab, the walls on the other building sides) could create indoor overheating in summer for the difficulties to dissipate the heat trapped.

The following subsection describes the two passive walls types (solar walls and Trombe wall) and reports experimental comparisons among different uses concerning shading and ventilation. Moreover, it includes numerical extensions to verify the impact of the adoption of the passive systems on the current superinsulated building envelopes.

Solar wall (Fig. 5.38A) is a passive solar system originally conceived for passive heating of building. It is generally made up of a not insulated south-facing massive wall painted black on the external surface, an air layer and glazing on the exterior side. It catches the solar radiation by means of the greenhouse effect created in the glazed cavity, absorbs and stores the thermal energy using the massive wall and finally exchanges it with the indoor environment by transmission through the wall. Shading devices such as overhangs or movable shutters provide solar radiation control.

Trombe wall (Fig. 5.38B) is a particular type of solar wall equipped with vents at the top and the bottom for the air thermocirculation between the air gap and the indoor environment. In a Trombe wall system heat exchange with the internal environment is both by transmission and by ventilation through the vents. Shading devices such as overhangs and roller shutters provide solar radiation control. The original usage instructions of the system (Fig. 5.38C) provided the following phases:

(i) *Winter—day*, a quote of the heat trapped in the interspace (open shutters) enters into the environment directly through the vents, another quote is stored by the mass;

(ii) *Winter—night*, after closing the adjustable vent and the outside roller shutter to limit heat dispersion, the wall gives back the heat accumulated by radiation;

(iii) *Summer—day*, the internal heat is stored by the mass (closed shutter);

(iv) *Summer—night*, the heat is dispersed toward the outside (open shutters) and the wall absorbs heat by radiation from the inside thus reducing its temperature.

The application of solar and Trombe walls is still problematic for a series of open problems regarding system management by the users, winter and summer performance and high environmental burdens. So optimization solutions are proposed hereafter.

a. *Problems regarding the system management*

The very complex original service schemes for the Trombe wall system management were its main drawback.

In *winter*, the original service drawings provided for the evening closure of the recirculation vents and shutters to be opened again in the morning. The closure of the shutters were intended to avoid the heat dispersion toward the sky, while the closure of the vents was to prevent the inversion of the thermal flow and the consequent introduction of cold air into the environment through the lower vents. The vents not only causes a complex operation for the users but also introduces the problem of the dust rising.

In *summer*, the service schemes provided to keep the vents always closed during night and day, thus contributing to a cooler inside environment. The shutters according to the initial schemes have to be closed during the day and open at night to allow the stored heat to be dispersed. However, some author presented unscreened version of the wall to enhance the solar chimney effect, introducing a still open debate on the most suitable screening use. To resolve these problems reduced operations and simplified system design are proposed hereafter (point d) and experiments on different screening use are presented (Section 5.3.2).

b. *Winter and summer performance*

Trombe wall has been widely studied regarding winter behavior, since this system was originally conceived for cold climates. In such period, the main drawbacks related to Trombe walls concern the uncertainty of ventilation exchanges and the inverse thermosiphon phenomena that can determine high heat losses [20]. Moreover a typical feature of solar walls is their low thermal resistance (to maximize the heat gains). Several improvements are proposed in literature to reduce winter losses: wall insulation and double glazing [21], use of a more complex design such as composite Trombe wall [22,23], deactivation of ventilation [24], and proper management [25].

Considering the summer season, the application of solar and Trombe walls can be the source of undesired heat gains and overheating phenomena [26]. These problems could become more severe in climates characterized by hot summer and in the cases in which the passive systems are adopted in a superinsulated and heat-tight building realized according to current energy saving standards. Actions to improve summer behavior can be grouped into three categories: ventilation, shadings, and insulation.

Ventilation of Trombe wall for summer cooling was studied by Gan [21] with a CFD numerical analysis. It was found that the ventilation rate induced by the buoyancy effect increases with the wall temperature, solar heat gain, wall height, thickness and insulation, distance between wall and glazing.

Solar shading of Trombe wall in summer is an action recommended by several authors [27] but only few of them determined its influence on the Trombe wall thermal behavior. Different types of shading are proposed in literature such as overhangs, shutters, and blinds. The behavior of a solar wall screened by overhangs was studied in hot climate conditions [28]. The experience proved that solar walls impose an additional cooling load on the building even if the system is screened by overhangs. Jaber and Ajib [29] recommend roller shutters and insulation curtains between glass and masonry wall layer for the optimum design of Trombe wall system in a Mediterranean region. Blasco Lucas et al. [30] experimented several passive systems, including a not ventilated solar wall, in comparison with reference systems. In summer the solar wall was screened by a PVC rolling curtain during the daytime and determined small heat gains over the traditional system. Several ways to prevent overheating with Trombe walls were compared by Ghrab-Morcos et al. [26] under hot climate conditions in Tunis. It was found that a very acceptable comfort level can be reached by screening the Trombe wall. The effect of a low-emissivity shading device in the air gap was studied in winter condition [31].

However, most of the abovementioned studies only give recommendations based on analytical simulations and rarely experimentally quantify the benefits of such solutions.

Summer overheating problems were resolved in the presented case study (see point d) with a combination of ventilation (through the introduction of ventilation dampers) and shadings (trough the adoption of roller shutters). The impact of such measures is quantified experimentally.

c. *High environmental burdens*

Finally, another open problem is the environmental impact of Trombe wall systems since they are often made up of materials and components characterized by high environmental burdens in the production phase. Moreover the overheating involves the use of a high environmental impacting electric energy. To give a contribution to this problem the environmental impacts of alternative Trombe walls solutions are quantified.

The case study: Simplified Trombe wall design and optimized summer performance.

The present section focuses on a Trombe wall in a solar residential building prototype in Ancona (central Italy) assumed as case study (case D.3 in Appendix A).

The system (Fig. 5.39) is made up of a 40-cm concrete wall painted black on the external side and plastered on the internal side. It presents a 10 cm thick air layer enclosed by a window with single glazing and

Figure 5.39 The solar house of Ancona.

aluminum framing. Adjustable vents are placed on the top (a) and on the bottom of the wall (b) to activate thermocirculation through the air gap. Roller shutters and their box are placed on the external side of the wall. The belts for roller shutters were placed inside black vertical aluminum bands insulated by means of polyurethane foam. Respect to traditional Trombe wall schemes, external openable dampers for summer cross-ventilation were adopted (see Table 5.7). They were designed as longitudinal bands inserted over the outside box (named "c" in the picture).

Thanks to previous research studies [32], an optimized method of control to solve the complex daytime and seasonal management was proposed.

In winter the purpose was to deactivate the air recirculation by maintaining all the vents closed and the shutter always open; the wall is anyway

Table 5.7 The optimized Trombe walls in the solar house of Ancona in summer, unshaded and shaded through the roller shutter

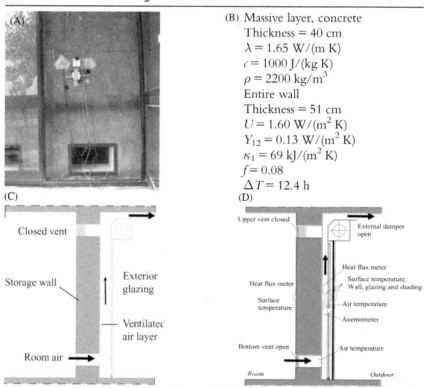

(B) Massive layer, concrete
Thickness = 40 cm
$\lambda = 1.65$ W/(m K)
$c = 1000$ J/(kg K)
$\rho = 2200$ kg/m^3
Entire wall
Thickness = 51 cm
$U = 1.60$ W/(m^2 K)
$Y_{12} = 0.13$ W/(m^2 K)
$\kappa_1 = 69$ kJ/(m^2 K)
$f = 0.08$
$\Delta T = 12.4$ h

(A) View of the concrete mass from the external side; (B) Thermal parameters; (C) Wall section; and (D) Monitoring instruments position.

able to store the heat during the day and to act as a thermal flywheel, behaving as a solar wall instead of a Trombe wall. Using the wall in this way, the upper vents are no more used (since the internal air recirculation is to be deactivated for the problem of dust) and could be removed from the system.

In summer, when the external upper dampers, the internal lower vents and the windows on the northern side of the apartment are open, hot air is drawn into the interspace. Comfortable cross-ventilation is thus created. Moreover in the hot period the shutters can be kept always closed (thus screening the capturing system), using the Trombe wall as a ventilated wall with opaque cladding (the PVC roller shutter), in which the inlet are located in the internal room rather than in the external environment.

Our research group carried out monitoring campaigns and numerical studies for several years to collect data regarding thermal behavior of solar walls and Trombe walls and the thermal comfort provided in different seasons [24,33]. Moreover starting from the as built case, numerical simulations in dynamic state allowed evaluating the influence of different wall materials and design features of solar wall on the environmental performances. Furthermore, a process of optimization was applied in order to determine the most suitable Trombe wall configurations by varying and combining several input parameters [34]. The influence of solar walls on the performances of the accommodation, at the increasing of the insulation level of the other building components (except the southern facade) was also evaluated [24].

5.3.2 Behavior of the Trombe wall, experimental measures

5.3.2.1 Behavior of the wall in summer, cloudy and sunny days

Trombe walls are convenient systems in Mediterranean climates in terms of energy performances and thermal comfort but proper measures should be taken to avoid the summer overheating drawbacks that could occur due to high heat gains. This section shows through measures that the adoption of external screening is more efficient than the maximum exploitation of the solar chimney by enhancing the buoyancy effect through high solar gains.

Figs. 5.40 and 5.41 reports the results of the summer monitoring of two Trombe walls realized in the southern exposure of the same room, but used in two different ways: one with closed shutters and the other in unscreened condition. Both walls were cooled by the ventilation of the

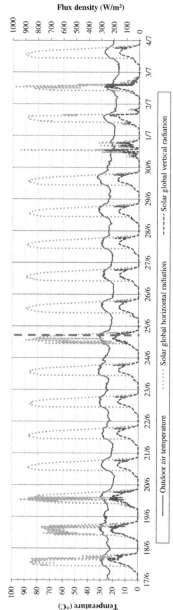

Figure 5.40 Weather data regarding wind recorded in the summer experimentation on Trombe walls.

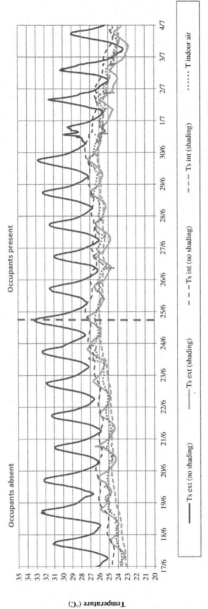

Figure 5.41 Summer experimentation. Internal and external surface temperatures of Trombe wall under different screening conditions.

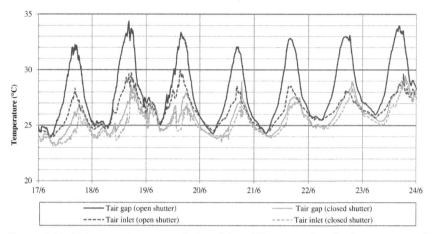

Figure 5.42 Air temperatures recorded in the middle gap and at the bottom vent of Trombe wall under different screening conditions. *From F. Stazi, A. Mastrucci, C. di Perna, Trombe wall management in summer conditions: an experimental study, Solar Energy 86 (9) (September 2012) 2839–2851, ISSN 0038-092X.*

external gap through the cross-ventilation with the openings facing to the northern side of the room. Fig. 5.40 reports the weather conditions of the monitored period. Fig. 5.41 shows that the closing of the external shutters reduces the internal surface temperatures of the wall (T_{sint}) of about 2°C. The air temperature inside the room is very close, and sometimes higher, than the surface temperature of the screened wall. So the wall can play its beneficial cooling effect.

Regarding the activation of the buoyancy force, it is clearly more effectively activated in the unscreened wall as demonstrated in Fig. 5.42. Indeed in this case there are higher temperature differences between the air in internal room (that is within the range of 23–25°C as seen in the previous figure) and in the air gap at middle height of the wall (ranging between 24 and 34°C in the unshaded wall and between 23°C and 29°C in the shaded wall).

5.3.2.2 Effect of the wind

Air velocity in the gap was compared in the case of unscreened and screened Trombe wall at the varying of wind velocity (Fig. 5.43). The wind speed probe was located adjacent to the wall to consider only the wind impacting frontally respect the Trombe system. Data collected demonstrated that the wind condition has an important influence on air velocity in the gap. On a windy day (June 19) the average air velocity in

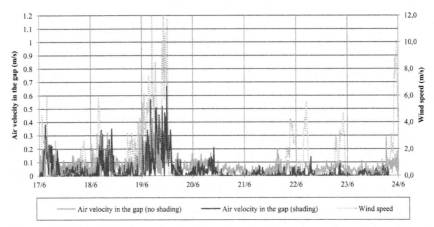

Figure 5.43 Air velocities recorded in the gap in screened and unscreened conditions.

the gap is similar in both screened and unscreened case, due to the influence of wind (0.12−0.13 m/s); on the day with low wind conditions (June 20), air velocities are lower, around 0.06 m/s without screening and 0.02 m/s with screening. The average air velocity for the period (June 17−23) was 0.06 m/s for the unscreened Trombe wall and 0.04 m/s for the screened one. Screening influences air velocity in the gap reducing the stack effect and the air flow rates. The effect is higher with low wind conditions.

5.3.2.3 Effect of the occupants' presence

The study of two summer days (June 21 and 27) with similar weather conditions and characterized respectively by the absence and presence of occupants (Fig. 5.44) shows that in the case of unscreened Trombe wall (A) the average air velocity is 0.06 m/s with occupation and 0.04 m/s without occupation. In the case of screened Trombe wall (B) the average air velocities are very low. However, this mode of use has a more beneficial effect in reducing heat gains: the heat fluxes (F_{int}) of the unscreened wall (C) are always entering in the room, being always positive regardless the occupants' presence. The presence of people determines an increase of the room air temperature not followed by a surface temperature increase (for the presence of a high mass) thus causing a reduction of the incoming heat. In the case of screened wall (D), the fluxes present a more variable behavior, being positive during the night and in the first hours of

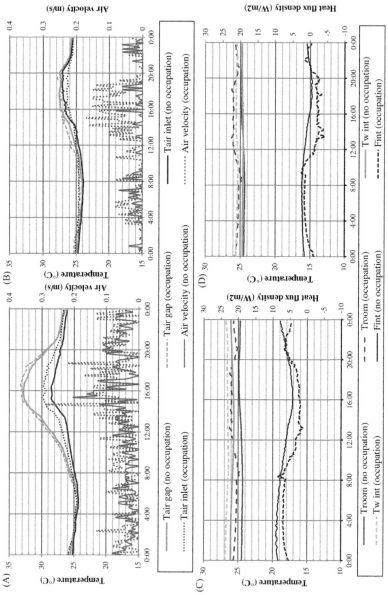

Figure 5.44 Summer experimentation. Influence of occupation on Trombe wall performance in both (A and C) unscreened and (B and D) screened conditions. *From F. Stazi, A. Mastrucci, C. di Perna, Trombe wall management in summer conditions: an experimental study, Solar Energy 86 (9) (September 2012) 2839—2851, ISSN 0038-092X.*

the day and negative in the afternoon thus allowing the wall to absorb the heat from the room. This behavior is even more visible if the occupants' are present.

The time lag of the massive wall was estimated to be 9 hours for the unscreened Trombe wall and 10 hours for the screened Trombe wall.

5.3.2.4 Behavior of the wall in intermediate seasons
In the intermediate seasons the heat gains are instead beneficial.

In October typical values of external temperatures range between 13°C and 23°C (Fig. 5.45). In such period the ventilation is deactivated, the roller shutters are maintained open, and the heating system is turned off. So the system is used as a simple solar wall. The heat stored during the summer determines surface temperatures always higher than the room temperature thus the heat flux is constantly directed from the wall to the room (Fig. 5.46). The heat flux is higher during the night when the differences between wall surface and air temperature are higher. This contributes to maintain a constant temperature of about 20°C, providing high comfort levels.

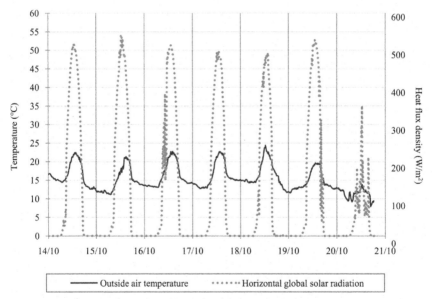

Figure 5.45 Results of the autumn monitoring on the Trombe wall accommodation. Outdoor conditions. *From F. Stazi, A. Mastrucci, C. di Perna, The behaviour of solar walls in residential buildings with different insulation levels: an experimental and numerical study, Energy Build. 47 (April 2012) 217–229, ISSN 0378-7788.*

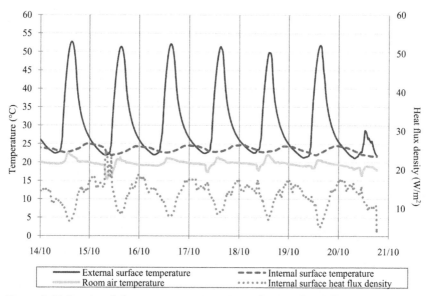

Figure 5.46 Results of the autumn monitoring on the Trombe wall accommodation. Indoor performance. *From F. Stazi, A. Mastrucci, C. di Perna, The behaviour of solar walls in residential buildings with different insulation levels: an experimental and numerical study, Energy Build. 47 (April 2012) 217−229, ISSN 0378-7788.*

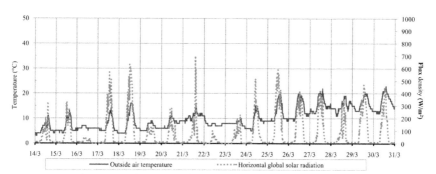

Figure 5.47 Weather data recorded in the March experimentation on Trombe walls. *From F. Stazi, A. Mastrucci, C. di Perna, The behaviour of solar walls in residential buildings with different insulation levels: an experimental and numerical study, Energy Build. 47 (April 2012) 217−229, ISSN 0378-7788.*

In March (Fig. 5.47) the environmental conditions are characterized by high values of solar radiation and variable temperatures: ranging between 5°C (at night) and 15°C in the first period (March 14−25) and between 10°C and 20°C in the last days of March. In all the period, the

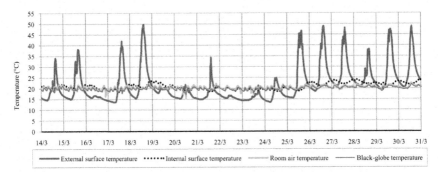

Figure 5.48 Extensive experimental measures on an intermediate season. *From F. Stazi, A. Mastrucci, C. di Perna, The behaviour of solar walls in residential buildings with different insulation levels: an experimental and numerical study, Energy Build. 47 (April 2012) 217—229, ISSN 0378-7788.*

ventilation of the wall was deactivated; the heating system was switched off the 25th of March (about 20 days before the typical turn-off day) thanks to the good environmental conditions guaranteed by the wall. In fact, in this period the internal surface temperatures were often higher than the values recorded for the air in the center of the room (Fig. 5.48) thus contributing with appreciable heat gains contribution.

5.3.2.5 Behavior of the wall in winter days

Results of winter monitoring are summarized in Figs. 5.49 and 5.50. Shading devices were open in the first part of the period (December 21—28) and closed in the second part to quantify their contribution to heat gain. The period was characterized by a combination of sunny and cloudy days. When the shadings are open, the solar wall external surface temperature gets up over 40°C on sunny days, while on cloudy days the surface temperatures are not far from outside air temperature. After a sunny day, the internal surface temperatures remain higher than room temperatures and the heat flux is directed toward the room (negative value), with daily heat gains of 0.45 MJ/m². On a cloudy day (December 25), the internal surface temperatures almost equal the room temperatures and this fact causes the inversion of the heat flux, from the room to the wall, and daily heat losses of 0.54 MJ/m². Closing the shading (after December 28), the solar wall thermal resistance is increased, but the solar radiation is excluded. In these conditions, the solar wall external surface temperature reaches 18°C when sunny, otherwise it maintains an average

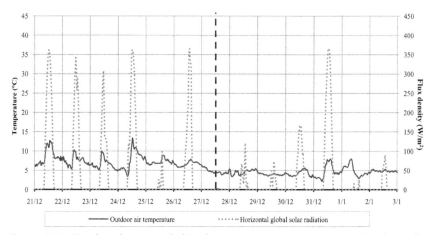

Figure 5.49 Weather data recorded in the winter experimentation on Trombe walls. *From F. Stazi, A. Mastrucci, C. di Perna, The behaviour of solar walls in residential buildings with different insulation levels: an experimental and numerical study, Energy Build. 47 (April 2012) 217–229, ISSN 0378-7788.*

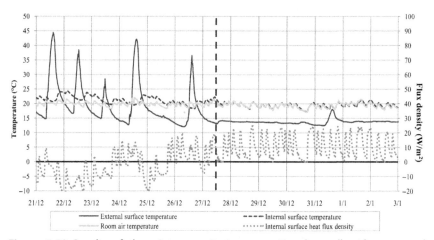

Figure 5.50 Results of the winter monitoring on a Trombe wall with open and closed shutter (after December 27). *From F. Stazi, A. Mastrucci, C. di Perna, The behaviour of solar walls in residential buildings with different insulation levels: an experimental and numerical study, Energy Build. 47 (April 2012) 217–229, ISSN 0378-7788.*

of 13.9°C. The internal surface temperature trend is very near to the inside air temperature trend. The heat flux is always directed from the room to the wall and the average daily heat loss is 0.98 MJ/m². It can be noticed that the sun radiation does not affect significantly the solar walls surface temperatures and heat fluxes in such conditions.

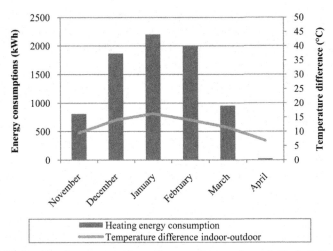

Figure 5.51 Results of energy consumption monitoring. *From F. Stazi, A. Mastrucci, C. di Perna, The behaviour of solar walls in residential buildings with different insulation levels: an experimental and numerical study, Energy Build. 47 (April 2012) 217–229, ISSN 0378-7788.*

Fig. 5.51 reports the monthly heating energy consumption and the difference between indoor and outdoor average air temperature obtained from the collection of data regarding gas consumptions of the accommodation according to UNI EN 15603:2008 over an entire year (see Section 6.4.2 for detailed method). Heating consumptions in central winter months are more than double the November and March energy demand, since in the latter months there are milder temperatures and higher exploitation of the passive heat gains.

5.3.3 Environmental sustainability

Solar walls are often made up of materials and components characterized by high environmental impacts in the production phase. This section reports the results obtained from energy analysis and LCA for the Trombe wall in the as built case, adopting as functional unit a portion of south-facing Trombe wall 2.50 m wide and 2.70 m high (surface 6.75 m^2). The environmental burdens were evaluated by considering the wall in unvented conditions, thus acting as a solar wall. The beneficial effect of the gap ventilation in summer is not considered in the cooling energy saving, while the roller shutter is considered closed. The service life of

Table 5.8 Functional unit of solar wall for life cycle assessment

| | Life span of facade systems | | | Building's life span |
	Plasters and paint	Sealants	Glazing	
Baseline LCA	25 years	25 years	50 years	50 years

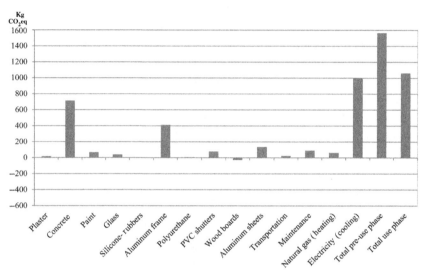

Figure 5.52 LCA results for the solar wall in the as built case using indicator IPCC 2001 GWP 100a. *From F. Stazi, A. Mastrucci, P. Munafò, Life cycle assessment approach for the optimization of sustainable building envelopes: an application on solar wall systems, Build. Environ. 58 (December 2012) 278–288, ISSN 0360-1323.*

the building was assumed to be 50 years. Maintenance operations consist in replacements of the system components at the end of their life. It is assumed that metal frames do not need replacement during the building service life. Service life of each component of the facades is reported in Table 5.8. The impacts relative to both preuse and use phase were assessed, including construction materials, transportation, and maintenance, natural gas for heating and electricity for cooling. The energy needs of the apartment were associated to the functional unit as explained in methods section (Section 6.6.12).

Results of the analysis using indicator IPCC 2001 GWP 100a (Fig. 5.52) showed that the Global Warming Potential is significantly high for the production phase of solar walls. In particular, relevant damages derive from the production of concrete and aluminum components. The other elements do not affect considerably the results as their impact is lower. Transportations do not determine overall high impacts. The highest impact in transportations is determined by the concrete due to its high mass; however, the short distance from the plant to the building site limits the burdens.

Moreover it is evident that the incidence of solar walls is much higher on cooling rather than on heating, with a percentage of 94% on the total impact for the use phase. This is mainly due not only to the high heat gains of solar walls in summer but also to the different impact of the two energy sources, since summer cooling is achieved by electrical power with low efficiency and winter heating by high efficiency fossil fuel.

5.3.4 Incidence of solar walls and Trombe walls at the varying of the building insulation level

This section deepens the efficacy of Trombe wall system at the increasing of the insulation level of the building in which the passive systems are adopted and demonstrates that such systems are still effective in overinsulated building provided that appropriate insulation level, design details, and management are selected.

Consider the envelopes shown in Table 5.9. Three types of building characterized by as many different insulation levels are considered:
- As built, adopting the envelope of the real case study (the solar house of Ancona), a prototype built up in 1983 with a low insulation level, as requested by the regulations of that period, in which the passive solar systems were commonly used;
- Conventional, with medium insulation level complying with the current regulations;
- Superinsulated, characterized by very low transmittance values for building components.

For each type of this building, the south-facing wall was varied among traditional wall (adopting W1, W2, and W3 according to the chosen insulation level), solar wall, and optimized solar wall. Solar wall is the same wall as in the as built case, while optimized solar wall is a type of solar wall improved with external double glazing. The incidence of screening solar walls with roller shutter (in both summer and winter) and the effect of system ventilation through the lower vent opening in summer were also evaluated.

Table 5.9 Building envelope insulation levels simulated

Vertical walls	As built	Conventional	Super-insulated	Southern exposure	
	Reference wall W1	Reference wall W2	Reference wall W3	As built solar wall	Optimized solar wall
Thickness (cm)	0.33	0.35	0.47	0.51	0.51
U (W/(m^2 K))	0.44	0.34	0.15	1.60	1.25
Y_{12} (W/(m^2 K))	0.07	0.04	0.01	0.13	0.08
κ_1 (kJ/(m^2 K))	49	49	48	69	68
f (−)	0.16	0.12	0.07	0.08	0.06
Time lag (h)	10.6	11.0	13.1	12.4	12.9
Glazing	Air 4-6-4 mm	Air, Low-E glass 4-12-4 mm	Argon, Low-E glass 6-12-6 mm	Single glass 4 mm	Double glass 6-12-6 mm
U (W/(m^2 K))	3.15	1.90	1.39	5.8	2.7

Horizontal slab (above garage)	As built	Conventional	Super-insulated
	Concrete–clay not insulated	With added insulation 14 cm	With added insulation 16 cm
Thickness (cm)	29	38	44
U (W/(m^2 K))	1.5	0.30	0.20
Y_{12} (W/(m^2 K))	1.15	0.09	0.06
κ_1 (kJ/(m^2 K))	38	39	39
$f(-)$	0.7	0.3	0.3
Time lag (h)	4.3	7.3	8.4

5.3.4.1 Is the solar wall convenient in superinsulated buildings?

The present subsection demonstrates that the adoption of a solar wall (unvented conditions) is not convenient in buildings with very high insulation levels.

Table 5.10 reports the comparison of seasonal heating and cooling energy needs for alternative southern facades (traditional, solar, and optimized solar wall) included in buildings with increasing insulation level (from low insulation as in the as built condition to superinsulated as the new trends).

Considering the as built and the conventional envelopes, it is possible to draw the same observations. The overall energy performance with solar walls (used with screening in summer and without screenings in winter) is better than with reference walls, achieving (e.g., in the as built case) a yearly energy need of 62.86 kWh/m^2 instead of 63.45 kWh/m^2. The superinsulation of the external envelopes (except for the southern facades) makes instead unfavorable the adoption of a solar wall on the southern exposure, determining adjunctive energy needs in both summer (for too high solar gains) and winter (for too low thermal resistance). The use of a solar wall with double glazing allows a further optimization of energy performance except for a building with superinsulation, in which a traditional wall also in the southern exposure should be preferred.

In summary the efficiency of solar walls (in comparison with a traditional one) decreases at the increasing of the envelope insulation level. The best energy saving amounts to −11.4% in comparison with reference case using optimized solar walls in the conventional envelope. In the case of superinsulated envelope the use of solar walls determines energy needs higher than the traditional walls.

5.3.4.2 Is the Trombe wall convenient in superinsulated buildings?

The present subsection demonstrates that the adoption of screenings and cross-ventilation on a solar wall, thus configuring a Trombe wall, is not enough to make this passive solar system convenient in buildings with very high insulation levels.

The importance of screenings in summer makes it interesting to quantify the system efficacy at the varying of envelope insulation level also in different shading conditions and with the cross-ventilation activated or deactivated. The unscreened condition (case A) is compared with alternative shadings: overhangs (case B), roller shutters (case C), and the

Table 5.10 Numerical simulation results

Type of building envelope	Reference wall	Solar wall		Optimized solar wall	
		Unscreened[a]	Screened[a]	Unscreened[a]	Screened[a]
As built envelope					
Seasonal heating energy needs (kWh/m^2)	60.56	**58.33**	68.35	**53.47**	66.91
Seasonal cooling energy needs (kWh/m^2)	2.89	9.19	**4.53**	10.82	**4.65**
Sum of seasonal heating and cooling energy needs (kWh/m^2)	**63.45**	62.86		58.12	
Conventional envelope					
Seasonal heating energy needs (kWh/m^2)	27.82	**25.52**	32.41	**20.91**	30.72
Seasonal cooling energy needs (kWh/m^2)	9.48	20.20	**11.69**	23.51	**12.13**
Sum of seasonal heating and cooling energy needs (kWh/m^2)	**37.30**	37.21		33.04	
Superinsulated envelope					
Seasonal heating energy needs (kWh/m^2)	13.48	**16.21**	21.52	**11.81**	19.80
Seasonal cooling energy needs (kWh/m^2)	10.53	23.31	**13.37**	27.17	**13.89**
Sum of seasonal heating and cooling energy needs (kWh/m^2)	**24.01**	29.58		25.70	

Seasonal heating and cooling energy needs of the accommodation at the changing of the envelope insulation level (as built, conventional, and superinsulated) and type of south-facing walls: reference walls, solar walls, (unvented) and optimized solar walls (unvented). The bold format regards the heating and cooling energy needs with the most convenient screening use (unscreened in winter, screened in summer) and the total energy needs by adopting these convenient modes of use.
[a]All the solution are simulated with upper overhangs as in the as built condition. The screened cases regard the closing of the external roller shutters.

combination of both overhangs and roller shutters (case D). All the cases are evaluated in both unvented and vented use of the wall. The results are shown, respectively, in Tables 5.11 and 5.12.

Results for unvented solar walls (Table 5.11) demonstrated that the use of shadings determines a decrease in cooling energy needs among −29.7% and −72.6% in comparison with the unscreened case depending on type of shading adopted and insulation level of the building envelope. Roller shutters (case C) are more effective than overhangs in reducing cooling loads, however, the best effect is obtained combining both types of shading (case D) with reductions among −59.7% and −72.6%. Moreover the reduction in cooling energy needs determined by solar wall's shading is more effective with a lower level insulation of the other elements of the building envelope. For instance, considering solar wall with both overhangs and roller shutters (case D), the reduction in cooling energy needs in comparison with the unscreened case is −72% for the as built envelope, −60.2% for the conventional envelope, and −59.7% for the superinsulated envelope.

Table 5.11 Numerical simulations results

Solar wall	(A) Unscreened	(B) Overhangs	(C) Roller shutters	(D) Overhangs + shutters
As built envelope				
Cooling energy needs (kWh/m^2)	16.52	9.19	6.08	4.53
Difference with unscreened case (%)	−	− 44.4	− 63.2	− 72.6
Conventional envelope				
Cooling energy needs (kWh/m^2)	29.40	20.20	12.28	11.69
Difference with unscreened case (%)	−	− 31.3	− 58.2	− 60.2
Superinsulated envelope				
Cooling energy needs (kWh/m^2)	33.15	23.31	13.92	13.37
Difference with unscreened case (%)	−	− 29.7	− 58.0	− 59.7

Summer cooling energy needs of the accommodation with unvented solar walls at the changing of type of screening for several types of building envelopes.

Table 5.12 Seasonal cooling energy needs of the accommodation with vented Trombe walls at the changing of type of screening for several types of building envelopes

Vented Trombe wall	(A) Unscreened	(B) Overhangs	(C) Roller shutters	(D) Overh. + shutt.
As built envelope				
Seasonal cooling energy needs (kWh/m^2)	14.29	8.32	5.81	4.48
Difference with the unvented case (same shading) (%)	− 13.5	− 9.5	− 8.2	− 1.3
Difference with the unscreened—unvented case (%)	− 13.5	− 49.7	− 64.8	− 72.9
Conventional envelope				
Seasonal cooling energy needs (kWh/m^2)	25.03	17.59	11.28	10.76
Difference with the unvented case (same shading) (%)	− 14.9	− 12.9	− 8.2	− 8.0
Difference with the unscreened—unvented case (%)	− 14.9	− 40.2	− 61.6	− 63.4
Superinsulated envelope				
Seasonal cooling energy needs (kWh/m^2)	28.13	20.18	12.66	12.26
Difference with the unvented case (same shading) (%)	− 15.1	− 13.4	− 9.0	− 8.3
Difference with the unscreened—unvented case (%)	− 15.1	− 39.1	− 61.8	− 63.0

The effect of cross-ventilation in combination with solar shading at the varying of building insulation level was also studied (Table 5.12). Firstly, comparison between the case of unvented and vented Trombe walls was made in terms of cooling energy need using equal shading conditions for every insulation level of the building envelope. This comparison highlights the benefit of activating ventilation of Trombe walls that determines a further reduction of cooling energy needs. The use of cross-ventilation for the case of Trombe walls without solar screening (case A) determines a reduction in cooling energy needs among −13.5% and −15.1% depending on the insulation level of the building envelope. This confirms that the single effect of ventilation is less important than the single effect of every type of solar screening (that was found above to be comprised between −29.7% and −72.6%). The reduction in cooling energy needs, due to the ventilation strategy, decreases at the increase of efficiency of shading system. For instance, in the as built envelope case, the reduction is −9.5% using overhangs (case B), −8.2% using roller shutters (case C), and −1.3% using a combination of overhangs and roller shutters (case D).

Secondly, the difference between each case and the case of unscreened−unvented Trombe wall, in terms of cooling energy need, highlights the overall benefit in using ventilation in combination with shading for every insulation level of the building envelope. The best result is obtained combining overhangs, roller shutters, and cross-ventilation (case D).

Moreover the reduction in cooling energy needs determined by Trombe wall's shading in vented condition is more effective with a lower insulation level of the other elements of the building envelope: the reduction in cooling energy needs, e.g., using overhangs and roller shutters (case D) is −72.9% for the as built envelope, −63.4% for the conventional envelope, and −63% for the superinsulated envelope. On the contrary, the reduction in cooling energy needs due to cross-ventilation of Trombe wall is more effective increasing the insulation level of the building envelope: the difference between vented and unvented case, using the same type of shading, is higher at the increase of the insulation level. For instance, using both overhangs and roller shutters (case D), the difference is −1.3% for the as built envelope, −8.0% for the conventional envelope, and −8.3% for the superinsulated envelope.

The data analysis reveals that even the adoption of proper measures such as screenings and ventilation is not enough to make the Trombe systems convenient in superinsulated buildings. As a matter of facts the yearly consumption for a superinsulated accommodation with traditional walls (W3) is 24.01 kWh/m^2 (see Table 5.10), while its amount for the same accommodation with Trombe walls on the southern side under optimal screening and ventilation conditions is of 24.07 kWh/m^2 (12.26 kWh/m^2 in summer and 11.81 kWh/m^2

in winter). This system is instead convenient in currently adopted insulation levels.

5.3.4.3 In summary, is the adoption of a Trombe wall still convenient today?

This passive solar system is still convenient with the insulation levels commonly adopted today, but a proper design regarding screenings and glazed surface should be adopted.

The study of total energy contribution of the wall to the environment was evaluated through the difference between solar gain and heat losses (Table 5.13). It demonstrates that adopting a structure with medium insulation level (conventional envelope) as common today, the choice of an optimized solar wall with double glazing is recommended in winter since it is the only wall that guarantees a net positive balance, with gains major than heat losses.

In winter (Fig. 5.53 left), the system is used as a solar wall, without ventilation and with screenings open. Reference (traditional) wall has a negative energy contribution for all months with heat losses higher than heat gains. Solar walls (with single glazing) present a worst behavior than traditional ones (for elevated heat losses) only for the coldest months. The adoption of double glasses in the optimized solar wall makes its behavior closer to a traditional (reference) wall, increasing its thermal resistance. Solar walls best performance is recorded during intermediate seasons when solar gains are high.

For the hot period, the total energy contribution was calculated considering a conditioned use of the accommodation and a screened use of the passive systems. The results for a building with conventional insulation level show that solar walls are mostly characterized by negative values of total energy contribution and show a better dissipating attitude than the reference case.

Table 5.13 Total energy contribution of reference walls and south-facing solar walls for different types of building envelope

Type of building envelope	Reference wall	Solar wall	Optimized solar wall
Conventional envelope			
Heating season total energy contribution (MJ/m^2)	−39.72	−35.46	+14.26

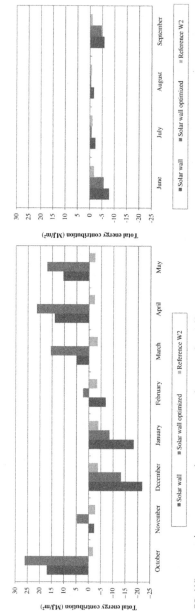

Figure 5.53 Winter and summer total energy contributions for different types of south-facing walls considering the accommodation with conventional envelope.

5.3.5 Optimal solution regarding system design and materials

The behavior of a solar wall is influenced by the type of massive layer (material and thickness) and by the thermal resistance of the external frame-glass system. This section deepens this aspect and identifies the optimal choice through the quantification of comfort levels and energy saving for different solutions.

Consider the envelopes shown in Table 5.14. Three different materials for the wall, characterized by unequal attitude to transfer/store heat are chosen: concrete is described by high conductivity and density (low thermal resistance, high storage ability), bricks are in a middle position, and aerated concrete is characterized by low conductivity and low density (high thermal resistance, low thermal capacity). Different thicknesses of the concrete massive layer were also taken into consideration (40, 30, and 20 cm).

The external glazed structure, consisting of framing and glazing, was also varied (Table 5.15), by considering three different technologies, aluminum, PVC, and wood for the frames and by increasing the thermal performance for the glazing, from single glasses to more performing types, such as low-e coated single glasses and double glasses.

Fig. 5.54 reports a graphical representation of LCA results for the construction phase (preuse phase, including raw material, material production, and transportation) and for the use phase (including operational energy associated to the functional unit). Moreover, to evaluate in detail the use phase burdens, the marginal energy needs for the walls alternatives are reported in Table 5.16. The marginal energy need represents the increase (or decrease) in energy need in kWh due to the presence of a solar wall on the south facade instead of a traditional (reference) wall. The detail methods for marginal energy needs calculations are reported in Section 6.6.1.2 (Step 2).

The results show that regarding the massive wall selection concrete and aerated concrete walls have the lowest impacts in the preuse phase. Brick wall has the highest impact for its energy-consuming production process. Regarding the use phase, the study of detailed energy data shows that the highest energy saving is guaranteed by adopting aerated concrete (-34.63 kWh/year) thanks to its lowest thermal transmittance value regardless the slight worsening of the decrementing attitude of the wall. This result indicates that the effect of reducing heat losses (for the reduced U value) predominates on the effect of reducing solar gains caused by a

Table 5.14 Alternative types of Trombe walls analytically evaluated

Wall	As built	Wall material		Wall thickness	
Massive layer	Concrete	Aerated concrete	Bricks	Concrete	Concrete
Thickness (cm)	40	40	40	30	20
Conductivity (W/m K)	1.65	0.17	0.68	1.65	1.65
Density (kg/m^3)	2200	500	1700	2200	2200
Specific heat (J/kg K)	1000	1000	940	1000	1000
Thickness (cm)	0.51	0.51	0.51	0.41	0.31
U (W/(m^2 K))	1.60	0.36	1.00	1.68	1.87
Y_{12} (W/(m^2 K))	0.13	0.04	0.06	0.24	0.48
κ_1 ((kJ/(m^2 K))	69	32	57	71	73
f (−)	0.08	0.1	0.06	0.14	0.25
Time lag (h)	12.4	15	15	9.8	7.3

Table 5.15 Alternative technologies evaluated for frame-glass system

Frames-glass	As built	Glass		Frames	
Description	Frames: aluminum Glazing: single	Single low-e	Double	PVC	Wood
U (W/(m² K))	2.4 frame	—	—	2.2	1.8
	5.8 glass	5.5	2.7	—	—

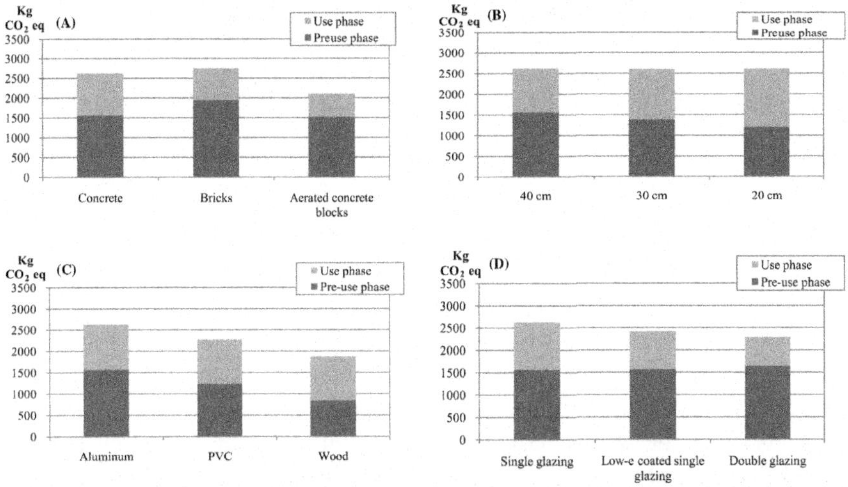

Figure 5.54 LCA results at the variation of the following parameters of solar wall: (A) wall material, (B) wall thickness, (C) frame material, and (D) glazing type. *From F. Stazi, A. Mastrucci, P. Munafò, Life cycle assessment approach for the optimization of sustainable building envelopes: an application on solar wall systems, Build. Environ. 58 (December 2012) 278–288, ISSN 0360-1323.*

higher thermal resistance. Differently cooling energy need is minimized with bricks. This depends on the minimization of decrement factor combined with a high κ_1 value. By an overall evaluation the aerated concrete blocks are the best choice since combine a production cycle with low environmental impacts and high energy performance.

The thickness decrease determines, as expected, an impact reduction in the preuse phase due to the minor material to be produced and transported and inversely an increase of environmental burdens in the use phase that could be explained with the increase of operating consumptions due to the smaller thermal resistance of the systems. The worst solution is those with concrete 20 cm thick since it has maximum both f and

Table 5.16 Marginal seasonal energy need for heating and cooling referred to the functional unit

Energy analysis	Marginal energy need for heating $Q_{h, fu}$ (kWh/year)[a]	Marginal energy need for cooling $Q_{c, fu}$ (kWh/year)[a]
As built	+4.89	+21.77
Bricks	−13.98	+21.28
Aerated concrete blocks	−34.63	+22.33
Concrete 30 cm	+14.77	+22.46
Concrete 20 cm	+27.62	+23.13
PVC	+3.29	+21.68
Wood	+3.09	+21.68
Low-e coated single glazing	−24.40	+25.27
Double glazing	−54.26	+29.15

[a]Refer to method section for the explanation of calculation methods for marginal energy need.

U values. However, from an overall evaluation the environmental impacts of the solar wall slightly decrease at the reduction of the wall thickness, due to the predominant influence of the preuse phase.

Regarding the frame-glass system, the high environmental burdens in the production phase connected to the aluminum window frames can be significantly reduced with the adoption of wood or PVC. In particular wood is very effective in terms of environmental performances due to a more sustainable production cycle. Also in the use phase the two latter frame materials with a higher thermal resistance show lower impacts for reduced operating heating and cooling consumptions. The wooden frame achieves the best outcome.

Changing the type of glazing determines a small increase in GWP in the preuse phase due to a higher impact of the production process. Inversely, considering the use phase the installation of more performing glazing determines lower heat losses and does not affect significantly the solar heat gains. By an overall evaluation, the strong decrease of operational energy using low-e single glazing and double glazing, determines a better environmental outcome than the initial single glazed solution.

The adoption of the optimized solution (Fig. 5.55), characterized by a 40 cm thick mass with aerated concrete, double glazing and wooden frames, allows to reduce the CO_2 emissions for both the production and use phases up to −55% respect to a traditional setup (concrete layer

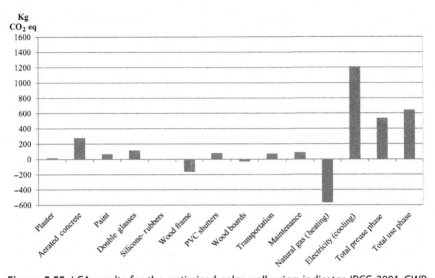

Figure 5.55 LCA results for the optimized solar wall using indicator IPCC 2001 GWP 100a. *From F. Stazi, A. Mastrucci, P. Munafò, Life cycle assessment approach for the optimization of sustainable building envelopes: an application on solar wall systems, Build. Environ. 58 (December 2012) 278–288, ISSN 0360-1323.*

40 cm thick, aluminum frame, and single glazing, see environmental impact in Fig. 5.52). However, the final choice should be based on different objectives, climates, and conditions set on the design process. For instance, in the case of unvented solar walls, a suitable choice could be the one of limiting summer energy needs (that resulted to be highly impacting) by adopting single glazing. In fact, the parameter that has demonstrated to have the highest influence on energy needs for cooling is the type of glazing (Table 5.16). Using single glazing instead of double glazing, the facade system will have lower energy needs for cooling and minimum GWP for the preuse phase.

5.4 DESIGN PATTERNS

Regarding the ventilated facades the choice of the external cladding and of the materials lying adjacent to the ventilation channel is very important determining a very different behavior of the walls. Two main behaviors are identified: conservative and selective.

Closed joint facades with an external cladding characterized by low thermal conductivity (as zinc-titanium braced by a "thermal insulating" wooden plank), and with a massive layer enclosed within the cavity, show a conservative behavior. The mass within the gap stores the heat during the day and releases it during the night thus determining a homogeneous and flattened behavior of the air temperatures within the cavity throughout the day and night.

Open joint facades with a massive cladding and with an insulation layer just placed behind the cavity (e.g., a clay-cladded wall) are selective solutions. The airflow is strongly influenced by the infiltration of external air and outdoor temperature fluctuations, with very different ventilation regimes between night and day.

As a consequence the activation of the chimney effect occurs at different hours for the two wall types: constantly for all the day and night but with low airflow rates in the metal cladded facade, only in the central day hours but with higher velocities in the clay-cladded wall. The presence of high wind regimes has effect on the zinc-titanium wall only when coming from its back, while on the open joint clay wall when hitting frontally. For both walls the higher is the wall, the major airflow is present. The ventilated walls show a good efficiency in southern building side but their adoption in eastern and western exposures is even more important since here the sun is more aggressive, striking with a low angle. In winter, the conservative titanium wall exhibits temperature values in the air gap always higher than the external temperature thus contributing to reduce the heating system working; differently for the clay wall the gap temperature equal the outdoor air temperature for almost all not sunny hours.

Hence, the conservative wall has a favorable behavior in winter, while the selective is more efficient in the hot periods.

The study of the environmental impacts demonstrates that the zinc-titanium walls have the greatest burdens and that the biggest weights are the construction materials, especially the bricks and the metal cladding, with the highest burdens on human health (Carcinogens) and on resources (Fossil fuels). For the clay-cladded wall, the highest total damage is caused by the clay blocks mainly weighting on Resource category due to fossil fuel large consumption for brick kilns.

The optimal ventilated facade is that with a ventilation air gap, interposed between a thick inner mass and an external insulation layer, which can be closed in winter through thermal performing vents. This solution shows optimal decrementing attitude in winter (very low f value), that is deactivated during the summer for the introduction of external air, thus allowing the dissipation of the heat stored in the internal mass. However, experimental data on this solution are still lacking.

Regarding the solar walls and Trombe walls, the summer monitoring revealed that the adoption of external screening is more convenient than the maximum exploitation of the stack effect as in the unshaded case. The best solution is obtained by combining overhangs, roller shutters, and cross-ventilation, using the facade as a cooling passive system that enhances through the chimney effect the natural cross-ventilation. In the intermediate seasons the solar heat gains are instead beneficial contributing to maintain a constant temperature of about 20°C. Results of winter monitoring show that the heat fluxes are always outgoing with high heat losses for closed shutters, while they present beneficial heat gains in the case of open shutters. As a consequence the system should be used with roller shutters open in winter and intermediate seasons and closed in summer. The air recirculation should be excluded for the dust rising, so the upper vents could be removed from the system.

The simulation highlighted that the Trombe walls could be conveniently adopted on the southern exposure of energy efficient buildings, designed according to the current regulations (with stationary transmittance U not lower than $0.30-0.35$ W/m^2 K). In a temperate climate they determine an energy saving of 24% on the sole winter heating consumptions, reduced to 11% on the total yearly consumptions for the slight increase of summer cooling energy needs.

Differently the adoption of Trombe wall systems in superinsulated buildings, designed according to the recent trends (with stationary transmittance U down to $0.15-0.20$ W/m^2 K) results in a total (winter + summer) energy increase ($+7\%$) for the problem of summer overheating.

The optimized solution for the system design concerns the adoption of 40 cm thick aerated concrete, with double glazing and wooden frames (with a thermal transmittance U value for the wall down to 0.36 W/m^2 K). Such wall, even if slightly increases the cooling consumptions, strongly reduces the winter energy needs thus resulting in reduced overall consumptions. Moreover, it lowers the environmental impact for the absence of highly impacting aluminum frames.

REFERENCES

[1] Directive 2010/31/EU of the European Parliament and of the Council of 19 May 2010 on the Energy Performance of Buildings. Official Journal of the European Union.

[2] Directive 2012/27/EU of the European Parliament and of the Council of 25 October 2012 on Energy Efficiency, Amending Directives 2009/125/EC and 2010/30/EU and Repealing Directives 2004/8/EC and 2006/32/EC. Official Journal of the European Union.

[3] C. Balocco, A simple model to study ventilated facades energy performance, Energy Build. 34 (2002) 469–475.

[4] P. Brunello, F. Peron, Modelli per l'analisi del comportamento fluidodinamico delle facciate ventilate, Fisica Tecnica Ambientale (1996) 313–324.

[5] M.J. Suárez, C. Sanjuan, A.J. Gutiérrez, J. Pistono, E. Blanco, Energy evaluation of an horizontal open joint ventilated façade, Appl. Therm. Eng. 37 (2012) 302–313.

[6] C. Balocco, Aspetti energetici delle facciate a ventilazione naturale in Facciate ventilate—Architettura, prestazioni e tecnologia, a cura di Frida Bazzocchi, Alinea editrice, 2002, pp. 269–326.

[7] F. PeciLópez, R.L. Jensen, P. Heiselberg, M. Ruiz de Adana Santiago, Experimental analysis and model validation of an opaque ventilated facade, Build. Environ. 56 (2012) 265–275.

[8] M. Ciampi, F. Leccese, G. Tuoni, Ventilated facades energy performance in summer cooling of buildings, Solar Energy 75 (2003) 491–502.

[9] F. Patania, A. Gagliano, F. Nocera, A. Ferlito, A. Galesi, Thermofluid-dynamic analysis of ventilated facades, Energy Build. 42 (2010) 1148–1155.

[10] P. Seferis, P. Strachan, A. Dimoudi, A. Androutsopoulos, Investigation of the performance of a ventilated wall, Energy Build. 43 (2011) 2167–2178.

[11] C. Sanjuan, M.J. Suárez, M. González, J. Pistono, E. Blanco, Energy performance of an open-joint ventilated façade compared with a conventional sealed cavity façade, Solar Energy 85 (2011) 1851–1863.

[12] E. Giancola, C. Sanjuan, E. Blanco, M.R. Heras, Experimental assessment and modelling of the performance of an open joint ventilated façade during actual operating conditions in Mediterranean climate, Energy Build. 54 (2012) 363–375.

[13] O. Zogou, H. Stapountzis, Energy analysis of an improved concept of integrated PV panels in an office building in central Greece, Appl. Energy 88 (2011) 853–866.

[14] W. Sun, J. Ji, C. Luo, W. He, Performance of PV-Trombe wall in winter correlated with south façade design, Appl. Energy 88 (2011) 224–231.

[15] A. Fallahi, Thermal performance of double-skin façade with thermal mass (Ph.D. thesis), Concordia University, 2009.

[16] F. Stazi, A. Vegliò, C. Di Perna, Experimental assessment of a zinc-titanium ventilated façade in a Mediterranean climate, Energy Build. 69 (2014) 525–534.

[17] F. Stazi, F. Tomassoni, A. Vegliò, C. Di Perna, Experimental evaluation of ventilated walls with an external clay cladding, Renewable Energy 36 (12) (2011) 3373–3385.

[18] R. Banham, The Architecture of the Well-Tempered Environment, Architectural Press, London, 1969.

[19] ASHRAE Handbook, Volume Fundamentals, American Society of Heating, Refrigerating and Air-Conditioning Engineers, Atlanta, 2005.

[20] H.Y. Chan, S.B. Riffat, J. Zhu, Review of passive solar heating and cooling technologies, Renewable Sustainable Energy Rev. 14 (2010) 781–789.

[21] G. Gan, A parametric study of Trombe walls for passive cooling of buildings, Energy Build. 27 (1998) 37–43.

[22] J. Shen, S. Lassue, L. Zalewski, D. Huang, Numerical study on the thermal behavior of classical or composite Trombe solar walls, Energy Build. 39 (2007) 962–974.

[23] L. Zalewski, M. Chantant, S. Lassue, B. Duthoit, Experimental thermal study of a solar wall of composite type, Energy Build. 25 (1) (1997) 7–18.

[24] F. Stazi, A. Mastrucci, C. Di Perna, The behaviour of solar walls in residential buildings with different insulation levels: an experimental and numerical study, Energy Build. 47 (2012) 217–229.

[25] A.V. Sebald, J.R. Clinton, F. Langenbacher, Performance effects of Trombe wall control strategies, Solar Energy 23 (1979) 479–487.

[26] N. Ghrab-Morcos, C. Bouden, R. Franchisseur, Overheating caused by passive solar elements in Tunis. Effectiveness of some way to prevent it, Renewable Energy 3 (6/7) (1993) 801–811.

[27] E. Tasdemiroglu, F.R. Berjano, D. Tinaut, The performance results of Trombe-wall passive systems under Aegean sea climatic conditions, Solar Energy 30 (2) (1993) 181–189.

[28] P. Torcellini, S. Pless, Trombe Walls in Low-Energy Buildings: Practical Experiences, Technical report, NREL Report No. CP-550-36277, National Renewable Energy Laboratory, 2004.

[29] S. Jaber, S. Ajib, Optimum design of Trombe wall system in Mediterranean region, Solar Energy 85 (2011) 1891–1898.

[30] I. Blasco Lucas, L. Hoesé, D. Pontoriero, Experimental study of passive systems thermal performance, Renewable Energy 19 (2000) 39–45.

[31] B. Chen, X. Chen, Y.H. Ding, X. Jia, Shading effects on the thermal performance of the Trombe wall air gap: an experimental study in Dalian, Renewable Energy 31 (2006) 1961–1971.

[32] F. Stazi, A solar prototype in a Mediterranean climate: reflections on project, use, results of the monitoring activities, calculations, in: Proceedings of the World Renewable Energy Congress (Innovation in Europe; Elsevier Ltd.) WREC 2005, 22-27 May 2005, Aberdeen, Scotland.

[33] F. Stazi, A. Mastrucci, C. Di Perna, Trombe wall management in summer conditions: an experimental study, Solar Energy 86 (9) (2012) 2839–2851.

[34] F. Stazi, A. Mastrucci, P. Munafò, Life cycle assessment approach for the optimization of sustainable building envelopes: an application on solar wall systems, Build. Environ. 58 (2012) 278–288.

CHAPTER SIX

Experimental Methods, Analytic Explorations, and Model Reliability

6.1 INTRODUCTION

The method adopted consists in an integrated approach that combines experimental and numerical steps for the optimization of energy saving, comfort, and environmental performance. The studies involved the following phases: a series of monitoring campaigns on several case studies, including a mock-up, in different seasons and usage conditions (regarding occupation, shadings, ventilation, heating system profiles); simulation in dynamic state on models calibrated with experimental data to evaluate alternative envelope solutions and different external climates and mode of use of the internal environment. Detailed information on the methods adopted for the experimentations and analyses are presented in this chapter.

6.2 THE GENERAL APPROACH

The present study is based on a multidisciplinary methodology involving the simultaneous analysis of different aspect through experimental and numerical phases:

1. Experimental on-site evaluation of the envelope performance:

    Thermographic survey to identify thermal bridges; on-site measure of the envelope thermal transmittance and comparison with the theoretical values; blower door test to quantify the actual envelope permeability.

2. Experimental study of the indoor comfort levels:

Thermal Inertia in Energy Efficient Building Envelopes.
DOI: http://dx.doi.org/10.1016/B978-0-12-813970-7.00006-6

Monitoring of indoor environmental conditions for rooms in various building levels and exposures in various season to check the comfort levels.

3. Energy analysis:

Simulations in dynamic conditions and parametric study with the software Energy Plus to evaluate the incidence of different retrofit strategies from the points of view of thermal comfort and energy saving; monitoring of real gas consumptions for some case study throughout the year.

4. Environmental sustainability and global costs: Application of Life Cycle Assessment (LCA) methods at the scale of both constructive element and building; global costs evaluation for different retrofit scenarios (also considering increasing insulation levels and various interventions combinations) in order to identify the optimal retrofit measures.

6.3 CASE STUDIES

The buildings used as case studies are all located in the Marche Region, in central Italy, characterized by a hot-summer Mediterranean climate (Köppen climatic classification). Most of the buildings are in the Adriatic coast, as in the cases built up in Ancona, or Porto Recanati while some of them are in the hilly part of the region, e.g., Passo di Treia or Villa Potenza.

The buildings were chosen as representative of different constructive typologies for the external envelope. For the ventilated walls two office buildings were selected but this type of system could be also adopted in multistory residential buildings. Moreover a school building was also monitored to consider a case with high internal heat gains.

A short description of each case study analyzed within the book is reported in Tables 6.1 and 6.2, respectively, regarding existing buildings and new or passive envelopes. A more detailed description of the same case studies and other new case studies are reported in Appendix A.

Table 6.1 Case studies evaluated for retrofit of existing envelopes

Traditional existing building envelopes	Single family		Multistoreys/apartment blocks	
	B.3	B.4	B.6	B.5
Year	1900–20	1945	1974	1981
City	Porto Recanati	Passo di Treia	Villa Potenza	Ancona
Lat./long.	43°27'/13°37'	43°17'/13°25'	43°19'/13°25'	43°35'/13°31'
Altitude on sea level	6 m	302 m	97 m	67 m
Interventions	First floor in 1920	–	Ext. insulation[a] in 1990	–
Orientation long. axis	45 degrees CW to N–S	NW–SE	NW–SE	N–S
S/V ratio	0.69	0.61	0.52/0.49	0.39
Window/wall ratio	7%	12%	14%	16%
External wall	Solid/semisolid bricks[c] $U=1.35/1.11$ W/(m² K)	Solid bricks $U=1.34$ W/(m² K)	Brick cavity wall $U=1.2$ W/(m² K)	Precast concrete panels $U=0.65$ W/(m² K)
Wall name	C2, C3	C1	S1	HR4
Roof	Brick-concrete slab $U=1.91$ W/(m² K)	Brick-concrete slab $U=2.74$ W/(m² K)	Brick-concrete slab $U=2.05$ W/(m² K)	Insulated brick-concrete slab $U=0.71$ W/(m² K)

(Continued)

Table 6.1 (Continued)

	Single family		Multistoreys/apartment blocks	
	B.3	**B.4**	**B.6**	**B5**
Traditional existing building envelopes				
Ground floor	Concrete slab laid above the ground level $U = 1.44$ W/(m^2 K)	Uninsulated brick–concrete slab $U = 0.91$ W/(m^2 K)	Brick–concrete slab low insulation $U = 0.72$ W/(m^2 K)	Insulated brick–concrete slab $U = 0.56$ W/(m^2 K)
Discussed in	Sections 2.5 and 3.4	Sections 3.3 and 3.4	Sections 3.3 and 3.5	Sections 2.4 and 3.6

[a]A building located near Building B.6 has identical characteristics, but in 1990 it was retrofitted with external insulation layer 5 cm thick. So the two cases were simultaneously monitored to verify the effect of the insulation layer.

Table 6.2 Case studies evaluated for new and passive envelopes

New envelopes	Single family		Multistoreys	
	C.3	D.3	D.2	D.1
Year	2010	1983	2010	2000
City	Fermo	Ancona	Ancona	Ancona
Lat./Long.	45°15'/13°45'	43°35'/13°31'	43°35'/13°31'	43°35'/13°31'
Altitude on sea level	319 m	67 m	67 m	67 m
Orientation long. axis	Diagonal along N–S	E–W axis	Cylinder	13 degrees CW to E–W
S/V ratio	0.38	0.5	0.3	0.29
Window/wall ratio	Ground floor: 15%	South: 20%	30%	32% (on vent. portion)
External wall	Light framed envelope $U = 0.12$ W/(m^2 K)	Trombe walls $U = 1.60$ W/(m^2 K)	Zinc ventilated facade $U = 0.31$ W/(m^2 K)	Clay ventilated facade $U = 0.41$ W/(m^2 K)
Type of wall	Lightweight	Solar wall/Trombe wall	Conservative wall	Selective wall
Roof	Light framed envelope $U = 0.12$ W/(m^2 K)	Brick-concrete slab $U = 1.5$ W/(m^2 K)	Brick-concrete slab $U = 0.3$ W/(m^2 K)	Brick-concrete slab $U = 0.36$ W/(m^2 K)
Ground floor	Concrete slab laid above the ground level $U = 0.18$ W/(m^2 K)	Uninsulated brick-concrete slab $U = 1.5$ W/(m^2 K)	Brick-concrete slab medium insulation $U = 0.27$ W/(m^2 K)	Brick concrete-slab low insulation $U = 0.7$ W/(m^2 K)
Discussed in	Sections 4.3.2 and 4.3.5	Sections 5.3.2 and 5.3.5	Sections 5.2.2 and 5.26	Sections 5.2.3 and 5.2.6

6.4 EXPERIMENTAL METHODS

6.4.1 Envelope actual performance and indoor comfort levels

Thermographic surveys of the external envelope were carried out using an infrared thermocamera in order to verify the presence of thermal bridges and the general envelope conditions (presence of wetted areas or water infiltrations). In the surveyed buildings the heating system was kept continuously on during the data acquisition period with the internal temperature set-point at 20°C. This kind of survey made it possible to identify the undisturbed points of the wall for each case study where to locate the probes to measure the on–site thermal transmittance.

Endoscopic inspections were done using a flexible endoscope in order to identify the layers making up the walls, and to check the possible presence of not insulated areas for insulation compaction due to degradation phenomena.

On-field monitoring of the thermal transmittance values was carried out according to ISO 9869 [1]. Data were acquired on the envelope side facing to north every 10 minutes by using two surface temperature probes and a thermal flow meter on the inner side of the wall, two surface temperature probes on the outer side of the wall, and an acquisition data system. According to the standard the thermal transmittance was calculated by dividing the average heat flux by the average difference in temperature between the inside and outside of the building. The following conditions were verified to ensure most accurate results: (1) the difference in temperature between the inside and outside of the building is at least 5°C, (2) the selected period for transmittance calculations was cloudy rather than sunny; (3) the monitoring of heat flow and temperatures was carried out for more than 72 hours; and (4) a thermographic camera was used to secure the homogeneity of the building element.

Laboratory tests on thermal conductivity were done for envelopes with existing insulations on samples extracted on-site. Cylindrical samples of insulation material were extracted from the walls and their current thermal conductivity λ was measured with a device based on the heat flow meter method (UNI EN 12664) [2] keeping the sample in stationary conditions. During the test, the specimens were placed between a hot plate (20°C) and a cold plate (5°C) in order to keep a constant temperature gradient through them. The surface temperature and the heat flow at the center of the samples were measured for 24 hours. The results were then compared with those obtained from samples of new insulation material with similar characteristics taken as reference of the initial conditions.

The values of thermal conductivity measured in laboratory for new and the extracted samples (λ_{design}) were also compared with the values declared by the producers at the time of building construction ($\lambda_{declared}$). Moreover, the corrected values ($\lambda_{corrected}$) according to the UNI EN ISO 10456 [3] were analyzed. This standard establishes that to compare the values measured in laboratory with the same declared it is necessary to correct the former with factors based on the temperature and the moisture content of the samples during the measuring. The obtained results were finally compared with the *m* parameter introduced by UNI 10351 [4], representing the thermal conductivity decrease in the average use conditions of the insulation materials. The eventual reduction of performance was finally ascribed to the greater moisture content in the extracted samples for the higher water absorption of the material caused by its partial degradation.

The measure of air tightness for the monitored apartments was done using the Blower Door Test (UNI EN13829:2002) [5]. According to the standard a pressure difference between the internal and external environments was imposed and the necessary airflow value to maintain it was detected. Thus the value of airflow rate (m^3/h) at a specific reference pressure (50 Pa) was determined. Then the air permeability of the envelope has been evaluated by calculating n50 (EN 13790 [6]). Test results were also used as input data for the virtual models.

Envelope dynamic performance was assessed according to ISO 7726:2002 [7] by recording the following parameters at the center of the wall or of the horizontal slabs facing outward (roof and ground floor slab): (1) the internal and external surface temperatures through a set of resistance temperature detection (RTD) sensors; (2) the incoming and outgoing heat flux through heat flux meters positioned on the internal side of the wall.

For the ventilated facades in addition to the abovementioned parameter the following data were also recorded: (3) detailed analysis of the thermophysical conditions at the inlet openings, at the ventilation channel mid-height and in the top part of the facade; (4) hot-sphere thermoanemometers to record the velocity and the temperature of the air in the ventilation channels at the three chimney heights (inlet, mid-height, top part of the facade).

For each case study the monitoring of the *indoor thermal comfort* in the rooms just adjacent to the external envelope was carried out according to ISO 7726:2002 [7] and included the following measurements:

- External environmental conditions, by using an external weather station with direct and global pyranomters, a combined sensor for the speed and the direction of the wind and a thermohygrometer with a double antiradiation screen.
- Indoor conditions, by means of indoor microclimate stations with globe thermometer probes, psychrometers, hot-wire anemometers and thermoresistances.

An example of the building plan with the positioning of the sensors is shown in Fig. 6.1 for a traditional envelope and in Fig. 6.2 for a ventilated one. Other examples of instruments positioning are reported in Appendix A.

The accuracy provided by the manufacturer for the probes is the following:

- Thermoresistances: tolerance according to IEC 751, accuracy 0.15°C (at 0°C).
- Heat flux meters: tolerance according to ISO 8302, sensitivity of 50 μV/Wm^2, accuracy 5% m. v./reading.
- Black globe temperature probe: 0.15°C (at 0°C).

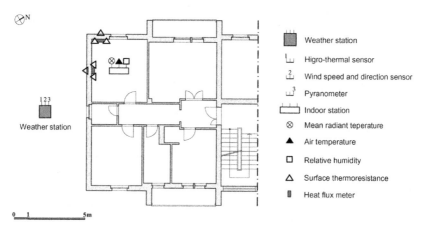

Figure 6.1 Measuring instruments in a traditional envelope. *From F. Stazi, A. Vegliò, C. Di Perna, P. Munafò, Experimental comparison between 3 different traditional wall constructions and dynamic simulations to identify optimal thermal insulation strategies, Energy Build. 60 (May 2013) 429–441, ISSN 0378-7788.*

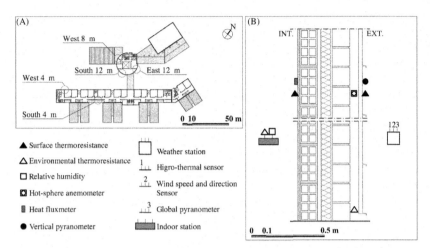

Figure 6.2 Measuring instruments in a ventilated envelope. (A) Building plan with the indication of the surveyed ventilated walls (various exposures and heights) and (B) section of the wall with the sensors positions. *From F. Stazi, A. Vegliò, C. Di Perna, Experimental assessment of a zinc-titanium ventilated façade in a Mediterranean climate, Energy Build. 69 (February 2014) 525−534, ISSN 0378-7788.*

- Thermohygrometer: temperature 0.15°C (at 0°C); UR 2% (5%−95%, 23°C).
- Hot-sphere anemometer: air velocity 0.03 m/s + 5% m. v./reading.
- Pyranometers: uncertainty < 2%.
- Wind direction probe: 5 degrees.
- Wind speed probe: 2.5% m. v./reading.
- Accuracy of data logger is 3% m. v./reading.
  Data takers DT500 were used with:
- voltage: resolution 1.3 µV; range ± 25 mV; tolerance ± 0.16% of full scale;
- RTDs, four-wire: resolution 0.01°C; range Pt100 (100 Ω); tolerance ± 0.17% of full scale;
- analogue to digital conversion: accuracy 0.15% of full scale; linearity 0.005%.

The monitoring included periods in which no changes were made to the habits of the occupants so as to verify the real use of the apartments and periods in which some particular use of the environment was imposed (e.g., activation of the natural ventilation, roller shutter always open or closed), changing only one condition at a time, to analyze its effect on the internal comfort level.

6.4.2 Consumptions

Data regarding gas consumptions of the accommodation were collected according to UNI EN 15603:2008 [8] over an entire year for some building case. The amount of methane delivered was monitored by gas meter reading. As it was not possible to have data for more than one entire year, correction for the weather was necessary. According to EN 15603:2008, energy content of gas delivered was determined multiplying the quantity of gas delivered by its gross calorific value.

Energy consumptions for other uses and for domestic hot water were detracted from total consumptions as gas is used both for heating and other uses. Reference values and calculation methods given by national standard UNI/TS 11300—2: 2008 [9] were applied for this scope.

6.5 ANALYTICAL METHODS

6.5.1 Dynamic simulation and data reliability

Dynamic thermal simulations with Energy Plus software [10] were performed to evaluate walls thermophysical parameters, internal comfort conditions, and energy consumptions.

The models were calibrated through comparison with monitored values. The real outdoor environmental conditions and the specific data of occupancy conditions (air infiltration, natural ventilation for windows opening, internal loads) were set on the models so the coincidence between the monitored and calculated values could be checked.

The detailed explanation of simulation and calibration methods is reported in Appendix B.

Fig. 6.3 reports an example (obtained in building 1) regarding the comparison of the walls surface temperatures at the ground level that shows that the model reproduces with a good approximation the real behavior.

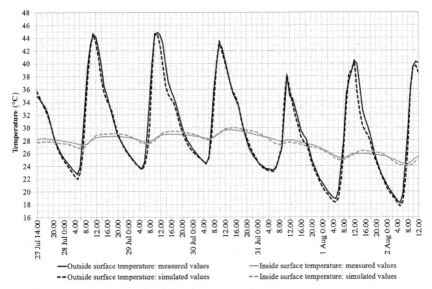

Figure 6.3 Example of calibration of the simulation model by comparison with the measured data. *From F. Stazi, C. Bonfigli, E. Tomassoni, C. Di Perna, P. Munafò, The effect of high thermal insulation on high thermal mass: is the dynamic behaviour of traditional envelopes in Mediterranean climates still possible?, Energy Build. 88 (February 1, 2015) 367−383, ISSN 0378-7788.*

6.5.2 Evaluation of indoor comfort levels through existing comfort models

In recent decades there has been a lot of research on thermal comfort leading to the development of two different approaches: the thermophysiological and adaptive. The most popular evaluation method for the first approach is based on the PMV model, introduced by Fanger [11] and developed in the subsequent EN ISO 7730 [12]. It is a "stationary" method which takes into account thermophysiological property of man and his thermal balance with the environment, including not only the internal thermal conditions (temperature, humidity, air movement) but also the type of clothing and the level of activity. It has been shown that this model approximates well the reality for buildings where the continuous use of the heating−cooling plants with a fixed set point determines nearly steady state conditions, while it is less appropriate for buildings with natural ventilation, characterized by a closer relationship with the outdoor climate and by occupants' actions for environmental control [13].

In the early 1970 Nicol and Humphreys [14] introduced the theory of adaptive comfort. However, only from the second half of the 1990s the theory was widely diffused. This theory stated that if the occupants are left free to adapt to the environment, e.g., changing clothing or opening the windows, then they can tolerate environmental conditions outside of the range recommended by the stationary theories. So the comfort temperature becomes function of the interaction between the subject and the environment. Later the same authors [15] have intensified their studies highlighting how the comfort mainly depends on the external environment, and have proposed a model for buildings in free-running mode (natural ventilation) in which the comfort temperature is obtained as a direct function of average monthly external temperature. This model also applies to the case of heated or cooled buildings, in which the optimum temperature coincides with the set-point temperature of the plants and this set point could be varied, according to the outside temperature with a slightly different formula than that for buildings with natural ventilation. The adaptation to the changes in outside temperature consists in the continuous variation of the set-point temperature. This continuous variation of the set-point does not reduce the levels of interior comfort but allows energy saving. The authors assume that the plant is switched on when the running mean outdoor temperature (T_{rm}) is lower than $10°C$ and therefore have proposed the following two equations to derive the temperature of comfort T_{comf}:

$$\text{For } T_{rm} > 10°C: \ T_{comf} = 0.33 T_{rm} + 18.8 \tag{6.1}$$

$$\text{For } T_{rm} < 10°C: \ T_{comf} = 0.09 T_{rm} + 22.6 \tag{6.2}$$

In the present book the comfort levels have been examined through the appropriate models during winter, summer, and intermediate seasons for all the case studies. A detailed analysis was also performed on some case study on the hottest rooms.

The winter comfort was evaluated with the Fanger's method through the calculation of the predicted mean vote index (PMV, ISO 7730) assuming the limits of the comfort zone provided for new buildings (-0.5 and $+0.5$ for category II) or for existing buildings (-1 and $+1$ for category III). The winter comfort was studied with both continuous and intermittent heating.

The summer comfort was assessed with the adaptive model based on EN 15251 [16] which, adopting the Humpreys' algorithm defines, for each building category (new or existing building), the upper and lower limit values of indoor operative temperature T according to the relation:

$$T_{max/min} = T_{comf} \pm 3 (\text{or} \pm 4)* \tag{6.3}$$

*depending on the selected building category (II or III).

The winter and summer comfort levels were then evaluated with the method of "Percentage outside the range" (Method A, Annex F, EN 15251), calculating the number and the percentage of occupied hours on the whole season in which the PMV (in winter) and the operative temperature (in summer) are outside the comfort range.

The summer comfort levels were also considered at the varying of natural ventilation profiles: a continuous profile set to 0.3 air changes per hour (ach) according to UNITS11300-1 [9]; a variable profile that provides 0.3 ach during the day (6 a.m.−6 p.m.) and 2.5 ach during the night (6 p.m.−6 a.m.) according to UNI 10375: 2011 [17].

Along the book the comparison between the two models PMV and adaptive, was done (see e.g., Table 4.8). In general it was found that the PMV model detects percentage of discomfort very higher than those obtained with the adaptive comfort model. The difference on the results depends on the range of conditions, in which people feel comfortable sensations, considered by the two comfort models in a different way. The Fanger's model defines a static (due to typical values of activity and clothing assumed) and narrow range of comfortable temperatures that lead to a greater discomfort computation. The model with adaptive approach defines a greater and more flexible range of comfort temperatures (strictly correlated with the outdoor temperature) that takes into account the capacity of building occupants to adapt themselves to the local and seasonal climate.

6.5.3 Energy performance through simulations in steady and dynamic regime

The energy performance of the buildings has been studied according to ISO 13790 and UNI/TS 11300 [6,9] in two phases: energy analysis in a semisteady state using *Termo* software with the aid of *Therm* software for the study of thermal bridges (according to ISO 14683 [18]); energy analysis in a dynamic state with *Energy Plus* software. The detailed explanation of simulation methods is reported in Appendix B.

The building consumptions were then evaluated in dynamic regime with different operating conditions of heating/cooling systems assuming a temperature set-point of 20°C for heating and of 26°C for cooling [9]. The introduction of a summer mechanical cooling system (as an alternative to the base scenario with natural ventilation) was assumed to find out summer consumptions.

Since summer cooling is often achieved by electrical power with low efficiency and winter heating by high efficiency fossil fuel, to make these two different forms of energy comparable, the consumptions were calculated in terms of primary energy by using two different conversion factors (1 for fossil fuel and 2.17 for electric energy, as defined by AEEG in EEN 3/08 [19]) and two appropriate coefficients of performance of the systems ($\eta = 0.8$ for heating and EER $= 2.5$ for cooling).

To compare stationary and dynamic results an initial analysis has been performed to ensure convergence between stationary and dynamic models by fixing the same input regarding geometry (net floor area and volume), opaque and transparent components transmittances, dispersions for ventilation (0.3 ach) and transmission, heating system (continuous operating mode), internal gains (4 W/m^2), and external contributions through glazed components. Subsequently additions (e.g., internal partitions, thermal bridges etc.) on the two models were realized (according to the different input contemplated in a semistationary or dynamic analysis) in order to make them closer to the real case study. In this way it was possible to compare the results obtained with these models, both complying with UNI/TS 11300.

Table 6.3 reports the results of a comparison between stationary and dynamic models regarding three techniques, characterized by the same stationary transmittance but with different dynamic properties (namely, W4, W5, and W6 detailed in Section 2.2.1, Table 2.1). Moreover in order to identify the relevance of the hourly climate variation (mainly due to the external solar radiation) on dynamic evaluations, the useful energy demands were also calculated with different assumptions of radiation contribution: without shading devices or considering a total shading of the opaque surfaces, by changing the assumptions on absorption coefficient of solar radiation and emissivity of the external surface materials, and also the closure of the shading devices in the glazed surfaces.

The results obtained in the steady state do not show a substantial difference in techniques classification as obtained in variable regime (about less than 1% in winter and 1.3% in summer). The stationary evaluations in unshaded conditions overestimate the consumption of about 60% respect to the dynamic assessment. The total exclusion of solar radiation on opaque and transparent surfaces shows a convergence between the values obtained with the two methods (with stationary values 36% higher).

These differences depend on the unequal assumptions that the two programs made in relation to some features: geometry, internal and external surface resistance of the building components, dispersion coefficients for ventilation, lost power by ventilation, external climate, lost energy for ventilation, thermal bridges, added only in the semistationary model with the aid

Table 6.3 Winter and summer consumption of buildings with different envelopes (see wall types in Table 2.1) obtained by semistationary and dynamic simulations with continuous operation for the city of Ancona, at the varying of assumptions for radiation contribution

Constr. techniques/analysis (kWh prim./m² year)	No shading devices		Totally shaded	
	Dynamic	Semistationary	Dynamic	Semistationary
Winter				
Masonry	22.68	36.32 (+60%)	30.24	41.34 (+36%)
Wood–cement	22.62	36.15	30.15	41.18
Wood	22.86	36.41	30.21	41.35
Summer				
Masonry	7.72	6.24	2.35	3.65
Wood–cement	7.80	6.32	2.35	3.70
Wood	7.92	6.32	2.63	3.70

Source: From F. Stazi, E. Tomassoni, C. Bonfigli, C. Di Perna, Energy, comfort and environmental assessment of different building envelope techniques in a Mediterranean climate with a hot dry summer, Appl. Energy 134 (December 1, 2014) 176–196, ISSN 0306-2619.

of *Therm* software. Moreover the steady procedure doesn't take into account the dynamic behavior of the thermal masses interacting with the fluctuations of the external and internal temperatures, since it doesn't work on an hourly basis but with fixed monthly values according to UNI 10349 [20].

6.6 ENVIRONMENTAL SUSTAINABILITY

6.6.1 Life cycle assessment

The LCA analysis has been performed according to ISO 14040 and ISO 14044 [21] with the following steps.

6.6.1.1 Step 1—goal and scope definition

The goal of this study is to optimize energy and environmental performances of the building envelopes. Optimization regards the minimization of the energy needs and environmental impacts due to the adopted envelope system in its life cycle by testing several setups.

The functional unit varies according to the specific scope: the whole building during its 75-year (or 50-year) service life when comparing buildings with completely different constructive techniques (see Section 4.4.4.), or a portion of the external facade when the optimization is focused only on a building envelope component (e.g., vertical passive walls, see Sections 5.2.5 and 5.3.3). The compared buildings (or component) have the same periods of durability and the same functional unit.

(A) System boundaries

Before starting the analysis, system boundaries were carefully established identifying unit processes and their own input and output data.

The unit processes examined in this study generally involves three life cycle phases: preuse, use, and end-of-life (recycling and disposal).

Preuse phase in this study includes: acquisition of raw material, production of material and components, transportation and in-site realization. Use phase considers the energy needs for maintenance, heating and cooling due to functional unit, during the life span.

The usual definition of phases in relation to the life cycle of buildings envisages the inclusion of maintenance operations in the use phase. In some cases, maintenance operations necessary for the envelope were taken into consideration as part of the preuse phase. The advantage of such an assumption is the ability to separate impacts related to construction materials and impacts related to thermal performance of the wall.

(B) Assumptions and limitations

For the study some assumptions and limitations have been done: some processes without incidence on the envelope construction techniques to be compared have not been considered, such as plants, foundations, basement, doors, and room partitions (considering only partitions separating apartments). The electricity for construction, maintenance operations, demolition and the transport of waste to recycling or disposal site were neglected since no reliable data are available and for their small relevance respect to total energy requirement [22–25].

The main processes involving the facade construction were taken into account. The postuse phase concerning recycling and disposal was in some cases omitted, as well as the building process, since other researchers demonstrated that it has a small contribution to the results when comparing different building components (less than 4%).

(C) Data quality

Software *SimaPro* [26] was used for LCA analysis. National Italian databases for inventory were not available. For this reason database Eco-

Invent (I) was used in order to obtain inventory data regarding material processing and activities during each phase of the assessment.

To quantify transportation distances from the production site to the building site, a market research was carried out. For each material or component, the extraction or production plant nearest to the building site was determined.

Heat and electricity demand for the use phase were derived from the simulations in transient state with software Energy Plus (see Appendix B for simulation methods).

6.6.1.2 Step 2—inventory analysis

An inventory of the environmental burdens associated with unit processes was produced. The life cycle inventory data used in the LCA study are obtained from Eco-invent v 2.01 databases.

(A) Preuse phase inputs

The preuse phase includes several processes concerning raw material extraction, material production, and transportations.

The amount of each material in terms of mass (kg) or volume (m^3) was computed in order to calculate the environmental loads of material extraction and production processes. The materials were obtained from the Eco-Invent database that was supplemented by data obtained from manufacturers interviews and literature [27,28]. In particular some materials have been modified taking a similar material and changing the weights or percentages of its subcomponents.

Transportation was assumed to be by a lorry of 16 ton and it was computed as the total fuel (diesel) consumption to carry the materials from production plant to building site.

The maintenance includes the replacement of certain materials and the relative distances to transport them. Maintenance operation was assumed as a replacement of the elements of the envelope at the end of their life span. Environmental burdens derive from raw material extraction, material production, transportations, and replacement operation.

(B) Use phase inputs

Inputs in the use phase include operational energy associated to the functional unit. Operational energy needs for both heating and cooling were calculated using a model in transient state and assumed to be constant over the adopted life span. The whole accommodation with the investigated envelopes was modeled using software Design Builder—Energy Plus on a yearly basis.

For the passive solutions (ventilated facades, solar walls), the energy needs of the functional unit were obtained from heating and cooling energy needs of the whole dwelling subsequently referred to the wall portion. Energy needs were calculated as the marginal seasonal energy need (for heating and cooling) of the functional unit. The marginal energy need represents the increase (or decrease) in energy need in kWh due to the presence of the passive wall on the south facade instead of a traditional wall.

The calculation was based on the comparison between energy needs of the real dwelling with passive walls on the south facade (and traditional walls on the other exposures) and the same dwelling with adiabatic walls instead of passive walls on the south facade. The use of adiabatic walls was experimented with good results by other authors [29]. The difference in energy needs between the two cases is attributable exclusively to the presence of the passive walls. The obtained value was then referred to the surface area of passive wall constituting the functional unit.

The complete procedure to obtain the marginal heating energy need of the functional unit $Q_{h,fu}$ (kWh) for the life span considered is described hereafter:

- Calculation of the yearly heating energy need of the accommodation with passive walls Q_h (kWh year) with the aid of simulation software;
- Calculation of the yearly heating energy need of the accommodation with adiabatic walls instead of passive walls $Q_{h,ref}$ (kWh year) with the aid of simulation software;
- Difference between the energy needs obtained in the two cases;
- Multiplication by a factor obtained as the ratio of surface of the functional unit (e.g., 1 m^2 of vertical wall) ($S_{passive,fu}$) to total vertical surface of passive walls ($S_{passive,tot}$);
- Multiplication by the number of years in the life span (n).

The following equation was used for the calculation:

$$Q_{h,fu} = (Q_h - Q_{h,ref}) \cdot \frac{S_{passive,fu}}{S_{passive,tot}} \cdot n \qquad (6.4)$$

Once the heating energy need of the functional unit had been obtained, the environmental burdens due to heating were calculated in *SimaPro* considering a boiler with nominal power 24 kWh and gas methane.

Marginal energy need for cooling due to the functional unit $(Q_{c,fu})$ was calculated with a similar methodology, considering the energy need of the accommodation (Q_c) and the energy need of the same accommodation with adiabatic walls instead of passive walls $(Q_{c,ref})$:

$$Q_{c,fu} = (Q_c - Q_{c,ref}) \cdot \frac{S_{passive,fu}}{S_{passive,tot}} \cdot n \qquad (6.5)$$

Environmental burdens due to electricity consumptions were computed considering the Italian national electricity supply.

(C) Disposal

Concerning to the disposal, it has been hypothesized the recycling of some materials as aluminum, glass, and steel, the incineration of the wood and the landfill for the remaining materials as gypsum and mineral wool.

6.6.1.3 Step 3—impact assessment

Life cycle impact assessment includes several mandatory stages, according to ISO 14040.

The first stage is the category definition, regarding the selection of impact categories, category indicators, and characterization models. The second is the classification stage, regarding sorting and assignment of Life Cycle Inventory parameters to the specific impact categories selected before. The third stage is the characterization: flows, that were previously categorized, are characterized according to the indicator selected and summed to measure the impact category total.

Normalization is a further, optional stage: the magnitude of category indicator is calculated relating to reference information. Finally results can be grouped, sorting the input categories, and weighted to get a single number for the total environmental impact.

The following indicators were used in the present analysis to describe the environmental impacts [30]:

- The Eco-Indicator 99 (EI 99). This indicator is widely used in the building sector since developed to simplify the interpretation and weighting of the results. It defines the environment with three types of damage, human health, ecosystem quality, and resources. Three perspectives are adopted as a way to deal with subjective choice on endpoint level: egalitarian, individualist, and hierarchist. The latter was chosen in the present evaluations.

- The Cumulative Energy Demand. It is an indicator of the energy use throughout the life cycle of goods or services. The analysis includes both the direct use of energy and the indirect (or gray) use due, for instance, to construction or raw materials. It represents an indicator of environmental impacts as far as the depletion of energy resources is concerned. On the basis of this indicator no statements on the environmental effects of products are possible since the effects of emissions and the use of non-energetic resources are neglected. Despite this limitation, this indicator is shown because the "gray energy" is widely used as an ecological indicator, especially in the building sector. Factors for characterization are given for several categories of energy resources nonrenewable (fossil; nuclear) and renewable (biomass; wind, solar, geothermal; water). Normalization is not applied for this method. In order to obtain a total energy demand value, the weighting factor for each category is equal to 1.
- The Environmental Priority Strategies (EPS 2000). This indicator, differently by the others, is based on an economical evaluation and so it has not the normalization. It attributes an economic value to the damage, considering the society willingness to pay in order to avoid a worsening of the considered condition or to find a remedial.
- The Global Warming Potential (GWP). It has been the focus of study of the scientific community because of its effects on climate change. On this regard, the level of equivalent emissions of CO_2 can be considered the most reliable key parameter. For that reason the indicator selected as exemplary for this analysis is IPCC 2001 GWP 100a, which gives an estimation of the GWP measuring the amount of heat trapped in the atmosphere by a greenhouse gas compared to the amount of heat trapped by a similar mass of carbon dioxide (standardized to a GWP of 1). The time horizon in this calculation was fixed to 100 years. Normalization and weighting are not included in this method.

The four methods used in this study have very different approaches (multicategory or single-issue) and assign a different weight to each category to translate the values into one impact indicator (Pt). For example, in "Human Health" category of Eco-indicator 99 and EPS 2000 methods, some values little appreciable in Characterization phase, become more relevant and comparable with other damage categories in Weighting analysis. However, this translation is strongly influenced by the subjectivity linked to the choice of different grouping and weighting methods (in this case, e.g., as regard Eco-indicator method, it was selected the weighting set belonging to the hierarchical perspective) [31]. As a consequence the adoption of multiple indicators is important to objectively compare different techniques.

6.6.1.4 Step 4—parametric analysis and interpretation of the results

Parametric analyses were performed assuming the as built case as reference and changing one input parameter at a time in the model. Such analyses are useful to compare the effects of different system configurations on energy use and environmental impacts, varying one parameter, while all the others are fixed. For each variation, energy simulation and LCA assessment were carried out. Significant environmental issues are highlighted and the optimal solution from the environmental aspect was identified.

6.6.2 Global costs

The economic analysis was done according to the procedure described in the UNI EN 15459 [32] by using the global cost methodology. The global cost is defined by the initial investment costs at the start of the measure; plus the present value of the sum of the running costs (e.g., fuel costs) during the calculation period; minus the net present value of the final value of components at the end of the calculation period. The global cost is directly linked to the duration of the calculation period τ and it can be written as:

$$C_G(\tau) = C_I + \sum_j \left[\sum_{i=0}^{\tau} C_{a,i} j \, R_d i - V_{f,\tau} j \right] \tag{6.6}$$

where $C_G(\tau)$ represents the global cost referred to starting year τ_0, C_I is the initial investment cost, $C_{a,i}(j)$ is the annual cost for component j at the year i (maintenance, replacement, and running costs), $R_d(i)$ is the discount rate for year i, $V_{f,\tau}(j)$ is the final value of component j at the end of the calculation period (referred to the starting year τ_0).

With regard to initial investment cost (C_I): the unit prices for products, including both furniture and application, were established from the current Italian pricelist. In particular the prices were derived from the DEI pricelist [33] for the buildings recovery, renovation, and maintenance. To evaluate the cost related to the innovative vented insulation solution, e.g., in Table 3.13 (see system description in Section 1.2.2), additional costs respect to a traditional external insulation layer were applied due to: deeper wall mechanical fasteners, additional insulating material and workmanship for the spacers supply and installation, electronic system for vents opening and expanded metal mesh that were preassembled in the special panel. The prices were obtained from market companies and considering the system as if it was industrially produced rather than handcrafted.

With regard to the annual costs for components (C_a) it consists of maintenance/replacement costs (C_m) and operation cost (C_o). For the maintenance costs, only those related to energy system were considered (2.75% of the investment costs related to heating and cooling systems). The timing for replacement of systems and building components was acquired from the Annex A of EN 15459 considering the same cost adopted for the initial investment. The operational costs for heating and cooling were obtained by multiplying the useful energy demands with the respective tariff (0.087€/kWh for natural gas and 0.2€/kWh for electricity after tax) [34].

The discount rate (R_d) was used to refer the costs to the starting year with the following relation:

$$R_d p = \frac{1}{1 + R_R/100}^{\text{P}}$$ (6.7)

where R_R is the real interest rate and p is the timing of the considered costs (i.e., number of years after the starting year).

The final value for each component (V_f) was determined by straight-line depreciation of the initial investment until the end of the calculation period and referred to the beginning of the calculation period.

Afterward the different cost components have been grouped into three categories: costs related to the building envelope, costs related to heating, and cost related to cooling.

REFERENCES

[1] ISO 9869-1:2014. Thermal Insulation—Building Elements—In-Situ Measurement of Thermal Resistance and Thermal Transmittance—Part 1: Heat Flow Meter Method.
[2] EN 12664:2001. Thermal Performance of Building Materials and Products Determination of Thermal Resistance by Means of Guarded Hot Plate and Heat Flow Meter Method—Dry and Moist Products of Medium and Low Thermal Resistance.
[3] ISO 10456:2007. Building Materials and Products—Hygrothermal Properties—Tabulated Design Values and Procedures for Determining Declared and Design Thermal Values.
[4] UNI 10351:2015. Materiali e prodotti per edilizia—Proprietà termoigrometriche—Procedura per la scelta dei valori di progetto.
[5] EN ISO 9972:2015. Thermal Performance of Buildings—Determination of Air Permeability of Buildings—Fan Pressurization Method.

[6] International Standard ISO 13790:2008. Energy Performance of Buildings—Calculation of Energy Use for Space Heating and Cooling.

[7] International Standard UNI EN ISO 7726:2002. Ergonomics of the Thermal Environment—instruments for Measuring Physical Quantities (European EN ISO 7726:2001).

[8] EN 15603:2008. Energy Performance of Buildings—Overall Energy Use and Definition of Energy Ratings.

[9] UNI/TS 11300-1, 2:2008. Energy Performance of Buildings. Part 1: Calculation of Building Energy Use for Space Heating and Cooling. Part 2: Calculation of Primary Energy Use and of System Efficiencies for Space Heating and for Domestic Hot Water Production.

[10] D.B. Crawley, C.O. Pedersen, L.K. Lawrie, F.C. Winkelmann, Energy Plus. Energy Simulation Program, ASHRAE J. 42 (2000) 49−56.

[11] P.O. Fanger, Thermal Comfort: Analysis and Applications in Environmental Engineering, McGraw-Hill, New York, NY, 1970.

[12] International Standard ISO 7730:2005. Ergonomics of the Thermal Environment—Analytical Determination and Interpretation of Thermal Comfort Using Calculation of the PMV and PPD Indices and Local Thermal Comfort Criteria.

[13] R. de Dear, G.S. Brager, Developing an Adaptive Model of Thermal Comfort and Preference, Center for the Built Environment, UC Berkeley, 1998.

[14] J.F. Nicol, M.A. Humphreys, Thermal comfort as a part of self-regulating system, in: Proceedings of the CIB Symposium on Thermal Comfort. Building Research Establishment, Watford, UK, 1972.

[15] J.F. Nicol, M.A. Humphreys, Adaptive thermal comfort and sustainable thermal standards for buildings, Energy Build. 34 (6) (2002) 563−572.

[16] European standard EN 15251:2007. Indoor Environmental Input Parameters for Design and Assessment of Energy Performance of Buildings Addressing Indoor Air Quality, Thermal Environment, Lighting and Acoustics.

[17] UNI 10375:2011. Calculation Method of Indoor Temperature of a Room in Warm Period.

[18] International standard ISO 14683. Thermal Bridges in Building Construction. Linear Thermal Transmittance. Simplified Methods and Default Values, 2007.

[19] Authority for Electric Energy and Gas (AEEG), EEN 3/08 deliberation. Conversion Factor of kWh in Petroleum Equivalent Tons Connected to Energy Efficiency Certificates, 2008.

[20] UNI 10349: 1994. Heating and Cooling of Buildings. Climatic Data.

[21] International standards ISO 14040, 14044. Environmental Management—Life Cycle Assessment—Principle and Framework, Requirements and Guidelines, 2006.

[22] G.A. Blengini, T. Di Carlo, The changing role of life cycle phases, subsystems and materials in the LCA of low energy buildings, Energy Build. 42 (2010) 869−880.

[23] L. Guardigli, F. Monari, M.A. Bragadin, Assessing environmental impact of green buildings through LCA methods: a comparison between reinforced concrete and wood structures in the European context, Proc. Eng. 21 (2011) 1199−1206.

[24] C. Scheuer, G.A. Keoleian, P. Reppe, Life cycle energy and environmental performance of a new university building: modeling challenges and design implications, Energy Build. 35 (2003) 1049−1064.

[25] D. Kellenberger, H. Althaus, Relevance of simplifications in LCA of building components, Build. Environ. 44 (2009) 818−825.

[26] SimaPro, Database Manual, Methods Library, PRé Consultants, 2008.

[27] G. Beccali, M. Cellura, M. Fontana, S. Longo, M. Mistretta, Analisi del ciclo di vita di un laterizio porizzato. La Termotecnica, 2009.

[28] B.V. Venkatarama Reddy, K.S. Jagadish, Embodied energy of common and alternative building materials and technologies, Energy Build. 35 (2001) 129−137.

[29] S. Citherlet, F. Di Guglielmo, J.B. Gay, Window and advanced glazing systems life cycle assessment, Energy Build. 32 (2000) 225—234.

[30] R. Frischknecht, N. Jungbluth, Implementation of Life Cycle Impact Assessment Methods. Final Report Ecoinvent 2000, Swiss Centre for LCI, Dübendorf, CH, 2003.

[31] P. Neri, Verso la valutazione ambientale degli edifici, life cycle assessment a supporto della progettazione eco-sostenibile, Alinea ed, Italy, 2008.

[32] International standard UNI EN 15459:2008, Thermal Performance of Buildings—Economic Evaluation Procedure for Energy Systems in Buildings.

[33] Prezzi informativi dell'edilizia: Recupero Ristrutturazione Manutenzione, DEI Tipografia del genio civile, 2013.

[34] AEEG Elaboration on Eurostat Data for Domestic Consumers, 2012.

GLOSSARY

Adaptive (approach) Physiological, psychological, or behavioral adjustment of building occupants to the interior thermal environment in order to avoid discomfort.

Characterization Compulsory LCIA stage where the categorized LCI flows are characterized, using one of many possible LCIA methodologies, into common equivalence units that are then summed to provide an overall impact category total.

Decrement factor f Ratio of the modulus of the periodic thermal transmittance to the steady-state thermal transmittance, $f = |Y_{12}|/U < 1$.

Discount factor Factor by which a future cash flow must be multiplied in order to obtain the present value.

Discount rate Definite value for comparison of the value of money at different times.

Functional unit Unit which defines what precisely is being studied and quantifies the service delivered by the product system, providing a reference to which the inputs and outputs can be related.

Global warming potential Indicator that measures the amount of heat trapped in the atmosphere by a certain mass of a greenhouse gas compared to the amount of heat trapped by a similar mass of carbon dioxide.

Grouping Optional LCIA stage that consists of sorting and possibly ranking the impact categories.

Internal areal heat capacity k_1 Parameter that describes the ability of a given component to store heat on the inner side.

Life cycle assessment (LCA) Compilation and evaluation of the inputs, outputs, and the potential environmental impacts of a product system throughout its life cycle.

Life cycle inventory (LCI) analysis Phase of life cycle assessment involving the compilation and quantification of inputs and outputs for a product throughout its life cycle.

Life cycle impact assessment (LCIA) Phase of life cycle assessment aimed at evaluating the significance of potential environmental impacts based on the LCI flow results.

Linear thermal transmittance Heat flow rate in the steady state divided by length and by the temperature difference between the environments on either side of a thermal bridge.

Normalization Optional LCIA stage where the results of the impact categories from the study are usually compared with the total impacts in the region of interest.

Operative temperature Uniform temperature of an imaginary black enclosure in which an occupant would exchange the same amount of heat by radiation and convection as in the actual nonuniform environment.

Periodic thermal transmittance Y_{12} Capacity of an opaque wall to time shift and to mitigate the thermal flux which crosses it over 24 hours.

Primary energy Energy that has not been subjected to any conversion or transformation process.

Set-point (of the internal) temperature Internal (minimum intended) temperature as fixed by the control system in normal heating mode, or internal (maximum intended) temperature as fixed by the control system in normal cooling mode.

Weighting Optional LCIA stage where the different environmental impacts are weighted relative to each other so that they can then be summed to get a single number for the total environmental impact.

Selected Examples of Buildings

I INTRODUCTION

The following 13 examples were reported as experimental evidence of the incidence of mass and superinsulation on the envelope performance in various conditions.

The selected case studies include a variety of buildings from single houses to multistory buildings, for residential and nonresidential use.

The examples show that the accurate choice of the envelope layers and internal linings could enhance its dynamic interaction with internal and external environment thus guaranteeing thermal comfort for occupants and energy saving.

A INFLUENCE OF HEAT CAPACITY ON ENVELOPE PERFORMANCE

A.1 Mock-up, Agugliano, Ancona, Italy

Table A.1 regards a test cell used for experimental purposes.

Table A.1 Mock-up with loadbearing cross laminated timber envelope

Cross-laminated envelope	1. Internal plasterboard (1.25 cm)	**Type of building**

Cross-laminated envelope
Wood

1. Internal plasterboard (1.25 cm)
2. Vapor membrane
2. Insulation rock wool (5 cm)
3. CLT (12 cm)
4. Insulation (10 cm)
5. External plaster (1 cm)

Thickness t = 30 cm
U = 0.22 W/(m² K)
Y_{12} = 0.017 W/(m²K)
κ_1 = 12 kJ/(m²K)
f = 0.076

Type of building
Experimental test room

Location
Agugliano, middle Italy, eastern coast (43°32′N; 13°23′E) 196 m above sea level

Climate
Hot summer Mediterranean climate Csa (Köppen classification) 2.064 degree-days

Year of completion
Built in 2013

Building design
Isolated volume, S/V = 1.8 (m⁻¹). It is a mock-up, namely, a full-scale test facility for the study of the hydrothermal behavior of building components under real outside climate conditions. The building is designed as a closed box without windows to avoid undesired and uncontrolled heat gains or losses. The HVAC system of the test facility is designed in such a way that both residential and fully conditioned indoor environment can be simulated, including heat gains due to greenhouse effect through windows

Building components
Foundation: cement floor slab. The upper floor slab is detached from the ground through a ventilated and insulated chamber
Exterior walls: cross-laminated envelope
Windows: no windows to avoid uncontrolled gains
Roof: wooden plank, EPS insulation (16 cm), OSB panel, waterproofing membrane
System: radiant heating on the ceiling with heat pump (to provide summer cooling) and electric radiators (to impose internal heat gains)

Experimental studies
Several monitoring campaigns to study the envelope dynamic behavior in different conditions:
1. A monitoring campaign regarded the comparison between different internal linings with increasing mass (κ_1 value)
2. Another study was focused on the simultaneous comparison of different ventilated walls (κ_2 values)
3. Another experimentation regarded the survey of a phase change material encapsulated in a massive board for internal finishing

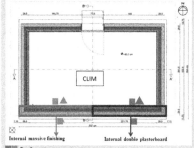

⊠ Internal massive finishing   Internal double plasterboard
▪▪ Surface temperatures
▲▲ Thermal fluxes
⊠ Ext. climate station
| CLIM | int. microclimate station

Realization phases

Table A.2 reports the mock-up realization phases. An accurate design allowed to strongly reduce the heat losses towards the ground and the air infiltration.

Table A.2 Details of the foundation system adopted, the cross laminated timber structure with the air barrier sheathing tapes and the internal finishings

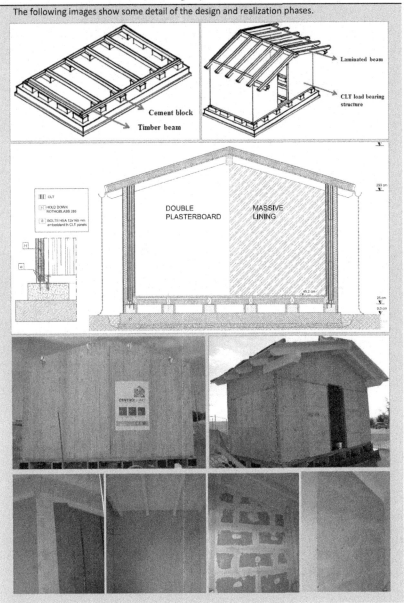

Comparison between envelopes at the varying of κ_1 value

Table A.3 shows preliminary data of a current experimentation on internal linings with increasing mass.

Table A.3 Schemes and preliminary data on the simultaneous measure of different internal linings

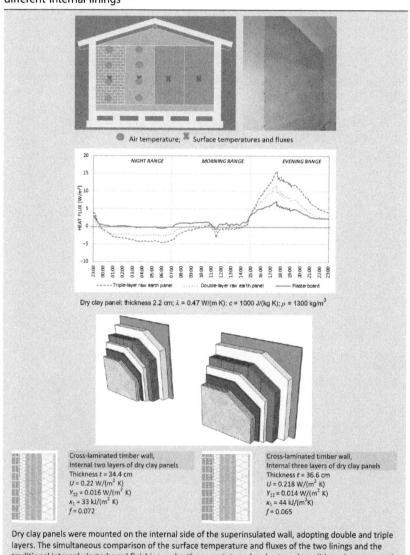

● Air temperature; ✖ Surface temperatures and fluxes

Dry clay panel: thickness 2.2 cm; λ = 0.47 W/(m K); c = 1000 J/(kg K); ρ = 1300 kg/m³

Cross-laminated timber wall,
Internal two layers of dry clay panels
Thickness t = 34.4 cm
U = 0.22 W/(m² K)
Y_{12} = 0.016 W/(m² K)
κ_1 = 33 kJ/(m² K)
f = 0.072

Cross-laminated timber wall,
Internal three layers of dry clay panels
Thickness t = 36.6 cm
U = 0.218 W/(m² K)
Y_{12} = 0.014 W/(m² K)
κ_1 = 44 kJ/(m²K)
f = 0.065

Dry clay panels were mounted on the internal side of the superinsulated wall, adopting double and triple layers. The simultaneous comparison of the surface temperature and fluxes of the two linings and the traditional internal plasterboard finishing under the same internal and external conditions demonstrated that both massive finishing (two and three layers) enhance the dynamic interaction of the envelope with the indoor environment. The triple-layer clay panel is the best choice since it stores the highest amount heat during the evening and release it at night.

Comparison between different ventilated envelopes

Table A.4 shows preliminary data of a current experimentation on ventilated envelopes at the varying of the mass position within the air gap.

Table A.4 Details and images of the experimented ventilated walls and preliminary data

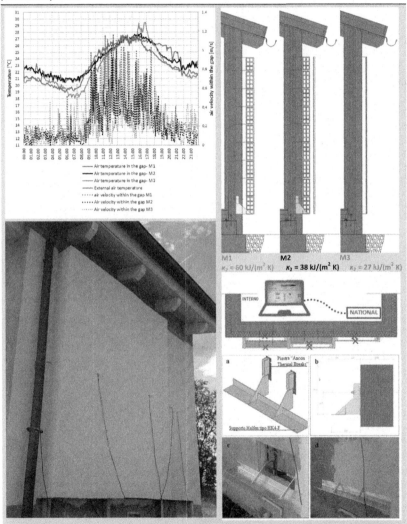

Three ventilated walls of different type were compared. M1 and M2 are characterized by the presence of a massive layer. In M1 it is the outermost layer, painted in white color; in M2 the ventilated channel is interposed between this massive layer and the external layer, with waterproof plasterboard. M3 is a ventilated wall without the massive layer, with external waterproof plasterboard. The wall with external mass M1 (high κ_2 value) is the best choice since in summer it reduces the incoming heat fluxes and cools the inner layers thanks to its more effective stack effect.

Phase change material

Table A.5 shows preliminary data of a current experimentation on phase change materials.

Table A.5 Details and images of the experimented clay boards enriched with microencapsulated phase change material and preliminary data

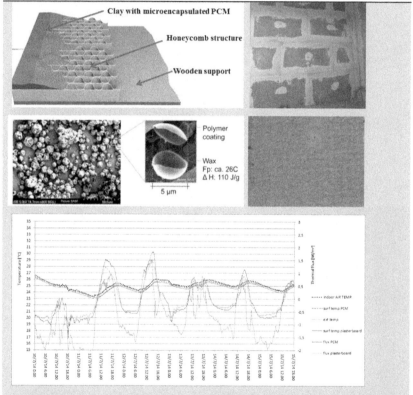

The adoption of phase change materials in the internal side of a superinsulated lightweight envelope enhances its dynamic interaction with the indoor environment. Heat is absorbed and released when the material changes from solid to liquid and vice versa. Thus it works as latent heat storage unit.

The comparison between the curves relative to heat fluxes of a traditional internal plasterboard and of the PCM material highlights the attitude of the latter to store the heat in the central hours of the day (higher positive fluxes) and to release it indoor during the night (higher negative fluxes). However, it is very important to choose the changing phase material accurately, since each panel works within specific temperature ranges. In the present study the paraffin wax with a phase change temperature of 22–23°C was selected, on the basis of preliminary internal temperatures surveys.

B RETROFIT OF EXISTING ENVELOPES

B.1 Residential House, Chiaravalle, Ancona, Italy

Table A.6 regards a monitoring of an isolated volume to deepen the effect of superinsulation on thick masses.

Table A.6 Single family house with superinsulated solid brick masonry envelope

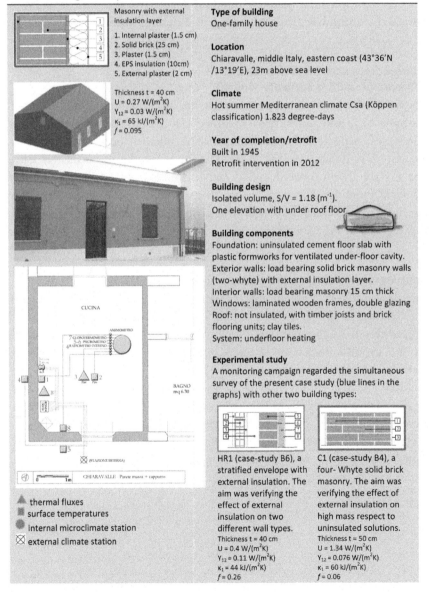

Masonry with external insulation layer

1. Internal plaster (1.5 cm)
2. Solid brick (25 cm)
3. Plaster (1.5 cm)
4. EPS insulation (10cm)
5. External plaster (2 cm)

Thickness t = 40 cm
U = 0.27 W/(m²K)
Y_{12} = 0.03 W/(m²K)
κ_1 = 65 kJ/(m²K)
f = 0.095

CUCINA

BAGNO
mq 6.30

7 CLOROTERMOMETRO
5-6 PSICROMETRO
4 RADIOMETRO INTERNO

ANEMOMETRO

(STAZIONE ESTERNA)

0   1m   CHIARAVALLE Parete massa + cappotto

▲ thermal fluxes
■ surface temperatures
● internal microclimate station
⊠ external climate station

Type of building
One-family house

Location
Chiaravalle, middle Italy, eastern coast (43°36'N /13°19'E), 23m above sea level

Climate
Hot summer Mediterranean climate Csa (Köppen classification) 1.823 degree-days

Year of completion/retrofit
Built in 1945
Retrofit intervention in 2012

Building design
Isolated volume, S/V = 1.18 (m⁻¹).
One elevation with under roof floor

Building components
Foundation: uninsulated cement floor slab with plastic formworks for ventilated under-floor cavity.
Exterior walls: load bearing solid brick masonry walls (two-whyte) with external insulation layer.
Interior walls: load bearing masonry 15 cm thick
Windows: laminated wooden frames, double glazing
Roof: not insulated, with timber joists and brick flooring units; clay tiles.
System: underfloor heating

Experimental study
A monitoring campaign regarded the simultaneous survey of the present case study (blue lines in the graphs) with other two building types:

HR1 (case-study B6), a stratified envelope with external insulation. The aim was verifying the effect of external insulation on two different wall types.
Thickness t = 40 cm
U = 0.4 W/(m²K)
Y_{12} = 0.11 W/(m²K)
κ_1 = 44 kJ/(m²K)
f = 0.26

C1 (case-study B4), a four- Whyte solid brick masonry. The aim was verifying the effect of external insulation on high mass respect to uninsulated solutions.
Thickness t = 50 cm
U = 1.34 W/(m²K)
Y_{12} = 0.076 W/(m²K)
κ_1 = 60 kJ/(m²K)
f = 0.06

Intermediate season performance: comparison with the stratified wall HR1

The stratified wall (HR1) presents higher outgoing (positive) fluxes than the massive insulated solution (C_{ext}) for its lower thermal resistance (Fig. A.1). The massive wall (C_{ext}) shows also a better interaction with the indoor environment thanks to its higher κ_1 value: the outgoing fluxes are partly accumulated within the mass during the day. During the night the heat is released inside (negative fluxes). The lower operative temperatures recorded in such masonry building depend on the great heat losses toward the ground as the graph in Fig. A.3 demonstrates.

Intermediate season performance: comparison with a capacity wall C1

The internal operative temperatures in the two buildings with insulated masonry C_{ext} and uninsulated masonry C1 are similar with maximum values occurring at different times, for the unequal occupants' habits in the system switching on (Fig. A.2). The fluxes for the not insulated solution C1 are always more than 5 W/m^2 higher than those recorded in the insulated building. The slightly lower operative temperatures recorded in the insulated building depend on the great heat losses toward the ground as the graph in Fig. A.3 demonstrates.

Heat losses influencing the final performance

The low indoor temperatures of the buildings with walls C_{ext} are mainly due to the great amount of heat losses toward the ground as demonstrated by the graph in Fig. A.3. On the left the fluxes recorded on the ground floor slab of such building are compared with those recorded in the building with HR1 envelope, on the right with the building having walls C1. The great contribution of this heat loss is clearly visible especially in the second period when the heating plants were switched on. This demonstrates that retrofit interventions concerning only the vertical envelope (C_{ext}) without insulating the horizontal ground floor and roof slab are not effective.

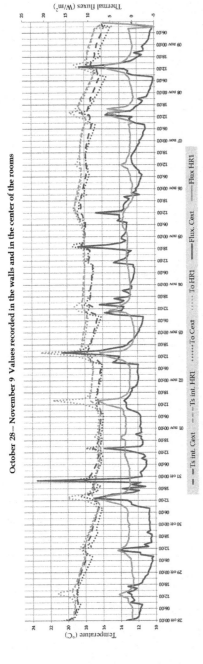

Figure A.1 Results of the monitoring campaign at the beginning of the cold season. Comparison of building B.1 with building B.6.

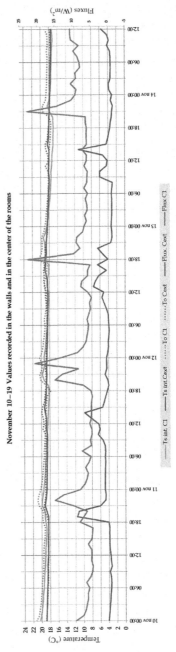

Figure A.2 Results of the monitoring campaign at the beginning of the cold season. Comparison of building B.1 with building B.4.

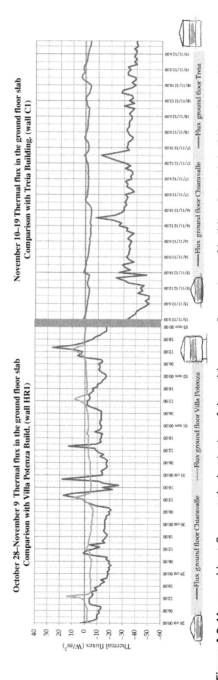

Figure A.3 Measured heat fluxes at the beginning of the cold season. Comparison of building B.1 with buildings B.6 and B.4.

B.2 Residential House, Jesi, Ancona, Italy

Table A.7 regards a monitoring on a compact residential volume to compare the behaviour of envelopes with different thermal inertia.

Table A.7 Two families house with solid brick masonry at the first two levels and hollow brick walls on the upper elevations

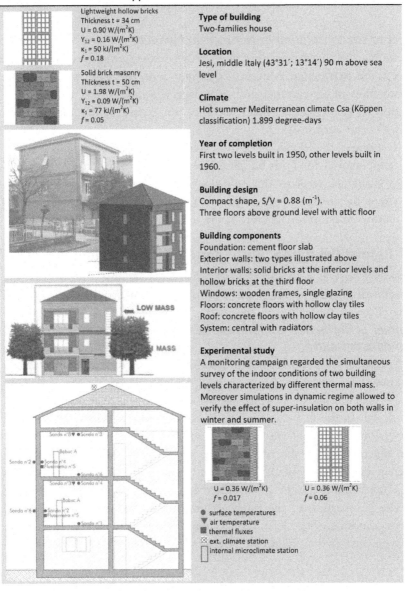

Lightweight hollow bricks
Thickness t = 34 cm
U = 0.90 W/(m²K)
Y_{12} = 0.16 W/(m²K)
K_1 = 50 kJ/(m²K)
f = 0.18

Solid brick masonry
Thickness t = 50 cm
U = 1.98 W/(m²K)
Y_{12} = 0.09 W/(m²K)
K_1 = 77 kJ/(m²K)
f = 0.05

LOW MASS

MASS

Type of building
Two-families house

Location
Jesi, middle Italy (43°31'; 13°14') 90 m above sea level

Climate
Hot summer Mediterranean climate Csa (Köppen classification) 1.899 degree-days

Year of completion
First two levels built in 1950, other levels built in 1960.

Building design
Compact shape, S/V = 0.88 (m⁻¹).
Three floors above ground level with attic floor

Building components
Foundation: cement floor slab
Exterior walls: two types illustrated above
Interior walls: solid bricks at the inferior levels and hollow bricks at the third floor
Windows: wooden frames, single glazing
Floors: concrete floors with hollow clay tiles
Roof: concrete floors with hollow clay tiles
System: central with radiators

Experimental study
A monitoring campaign regarded the simultaneous survey of the indoor conditions of two building levels characterized by different thermal mass. Moreover simulations in dynamic regime allowed to verify the effect of super-insulation on both walls in winter and summer.

U = 0.36 W/(m²K)
f = 0.017

U = 0.36 W/(m²K)
f = 0.06

● surface temperatures
▼ air temperature
■ thermal fluxes
⊠ ext. climate station
⌐ internal microclimate station

Intermediate season performance: comparison between levels with different mass

The experimental comparison of the surface temperatures recorded on the inner side of the external envelope at the two levels (Fig. A.4) shows that in the intermediate season (September 25–30), when the external temperature is still high (ranging between 16°C and 30°C), the massive wall maintains lower values with a flatter trend.

Effect of superinsulation in summer and winter

The numerical simulations (Fig. A.5) allowed comparing the not insulated walls at the two building levels (*dotted lines*) with the same walls after the introduction of an external insulation layer (*continuous lines* with massive solution in *blue* (black in print versions) and lightweight solution in *red* (gray in print versions)).

In summer, the insulation of the external walls reduces the internal surface temperatures for both building levels of about 1°C in very hot days, while it slightly worsens the behavior in not extreme days (August 1). In winter the insulation gives great benefices, with the best behavior of the massive solution for its more stable trend.

B.3 Residential House, Porto Recanati, Ancona, Italy

Table A.8 regards a monitoring of an isolated volume to compare the behavior of single layer envelopes with different thermal inertia.

Summer performance: comparison between levels with different mass

The graph in Fig. A.6 regards the comparison between two very similar days (see the external temperature trends in July 29 and 30, *blue lines* (light gray in print versions)) in which the windows were kept alternatively open and closed. In the first case the two building levels recorded equal internal temperatures (*black and gray dashed lines*). The window closing (*continuous lines*) determines that the two building levels greatly differ mainly for the different contribution of the other heat loosing or heat gaining elements (e.g., heat loss through the ground floor and heat gains through the roof).

The comparison of the internal surface temperatures with open and closed windows (Fig. A.7) reveals that there is a recurring difference between the two walls that could be ascribed to their thermal mass.

Table A.8 Single family house with solid brick masonry at the ground floor and hollow brick walls at the first floor

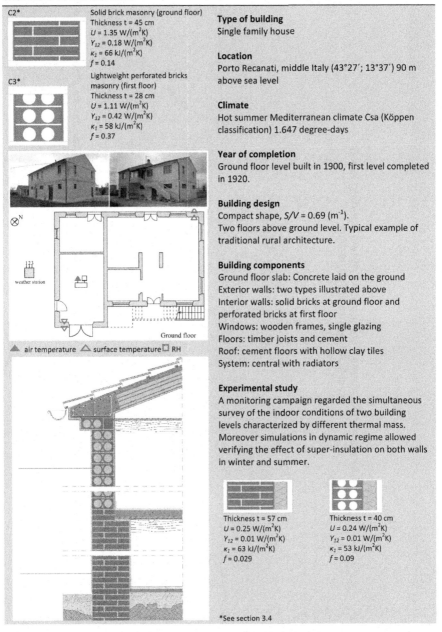

C2*	Solid brick masonry (ground floor) Thickness t = 45 cm $U = 1.35$ W/(m²K) $Y_{12} = 0.18$ W/(m²K) $\kappa_1 = 66$ kJ/(m²K) $f = 0.14$
C3*	Lightweight perforated bricks masonry (first floor) Thickness t = 28 cm $U = 1.11$ W/(m²K) $Y_{12} = 0.42$ W/(m²K) $\kappa_1 = 58$ kJ/(m²K) $f = 0.37$

Type of building
Single family house

Location
Porto Recanati, middle Italy (43°27′; 13°37′) 90 m above sea level

Climate
Hot summer Mediterranean climate Csa (Köppen classification) 1.647 degree-days

Year of completion
Ground floor level built in 1900, first level completed in 1920.

Building design
Compact shape, $S/V = 0.69$ (m⁻¹).
Two floors above ground level. Typical example of traditional rural architecture.

Building components
Ground floor slab: Concrete laid on the ground
Exterior walls: two types illustrated above
Interior walls: solid bricks at ground floor and perforated bricks at first floor
Windows: wooden frames, single glazing
Floors: timber joists and cement
Roof: cement floors with hollow clay tiles
System: central with radiators

Experimental study
A monitoring campaign regarded the simultaneous survey of the indoor conditions of two building levels characterized by different thermal mass. Moreover simulations in dynamic regime allowed verifying the effect of super-insulation on both walls in winter and summer.

N

weather station

Ground floor

▲ air temperature △ surface temperature ▢ RH

Thickness t = 57 cm
$U = 0.25$ W/(m²K)
$Y_{12} = 0.01$ W/(m²K)
$\kappa_1 = 63$ kJ/(m²K)
$f = 0.029$

Thickness t = 40 cm
$U = 0.24$ W/(m²K)
$Y_{12} = 0.01$ W/(m²K)
$\kappa_1 = 53$ kJ/(m²K)
$f = 0.09$

*See section 3.4

Source: From F. Stazi, C. Bonfigli, E. Tomassoni, C. Di Perna, P. Munafò, The effect of high thermal insulation on high thermal mass: is the dynamic behaviour of traditional envelopes in Mediterranean climates still possible?, Energy Build. 88 (February 1, 2015) 367–383, ISSN 0378-7788.

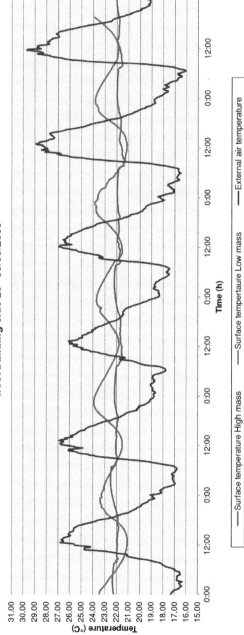

Figure A.4 Results of the monitoring campaign in the intermediate season. Comparison between two building levels with different mass.

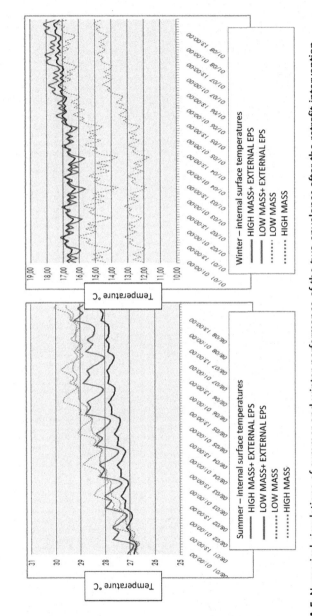

Figure A.5 Numerical simulations of summer and winter performance of the two envelopes after the retrofit intervention.

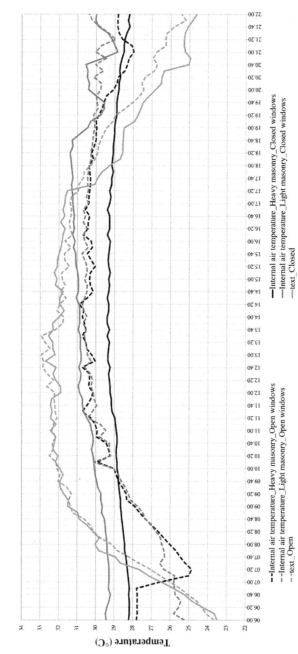

Figure A.6 Results of the summer monitoring campaign. The opening of the windows nullifies the differences between the two rooms.

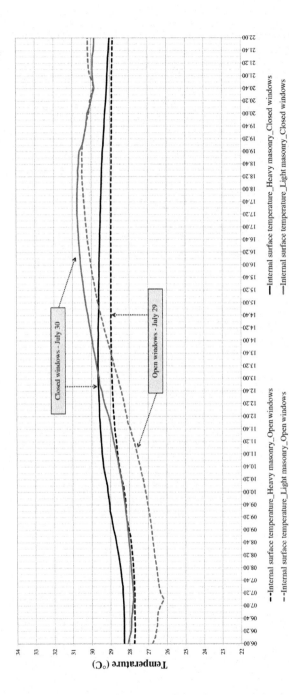

Figure A.7 Results of the summer monitoring campaign. Surface temperatures recorded at the two building levels under different conditions (windows open and closed).

B.4 Residential House, Passo di Treia, Ancona, Italy

Table A.9 regards the monitoring of an isolated volume to deepen the behavior of a very thick solid brick masonry and to make comparisons with other techniques.

Table A.9 Single family house with four-whyte solid masonry envelope

C1[a]

Solid brick masonry C1

1. Internal plaster (1.5 cm)
2. Solid brick (25 cm)
3. Plaster (1.5 cm)

Thickness t = 50 cm
U = 1.34 W/(m^2 K)
Y_{12} = 0.076 W/(m^2 K)
κ_1 = 60 kJ/(m^2 K)
f = 0.06

Type of building
One-family house owned by Social Housing Agency

Location
Chiaravalle, middle Italy, eastern coast
(43°17″N/13°25″E), 302 m above sea level

Climate
Hot summer Mediterranean climate Csa (Köppen classification) 1.823 degree-days

Year of completion/retrofit
Built in 1945

Building design
Compact shape, S/V = 0.61 (m^{-1}).
Two elevations with attic floor

Building components
Foundation: presence of a mezzanine floor
Exterior walls: load-bearing solid brick masonry walls
Floors: brick-concrete slab
Interior walls: load-bearing masonry
Windows: wooden frames, single glazing
Roof: pitched roof with brick concrete slab
System: central with radiators

Experimental study
A monitoring campaign regarded the yearly survey of an apartment (highlighted in *red* in the figure) Moreover the building was simultaneously monitored with case studies B1 and B6 to compare the behavior of different building envelope typologies. The comparison is reported in Section B1.

▲ Thermal fluxes
■ Surface temperatures
● Internal microclimate station
⊠ External climate station

[a]See Sections 3.3 and 3.4

Summer season performance

The internal surface temperatures recorded on summer (Fig. A.8) show a very flatter trend. The fluxes are always outgoing given the low external temperatures of the period in the selected climate characterized by high seasonal variations. So the mass cools itself and it is able to guarantee stable temperatures for the following hot days.

Winter season performance

In winter period the use of the heating plants guarantees comfortable indoor operative temperatures that fall within a range of 17−21°C (Fig. A.9).

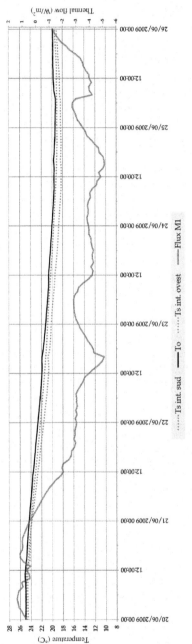

Figure A.8 Results of the monitoring campaign at the beginning of the summer.

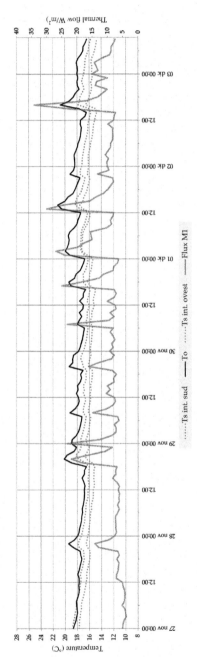

Figure A.9 Results of the winter monitoring campaign.

However, there is great heat loss, since the thermal fluxes are always outgoing with peaks corresponding to the system switching on.

B.5 Residential House, Ancona, Italy

Table A.10 regards the monitoring of two apartment blocks for the comparison of envelopes in different positions within the building volume and under unequal occupant's behaviour.

Table A.10 Social housing building with precast concrete load bearing panels

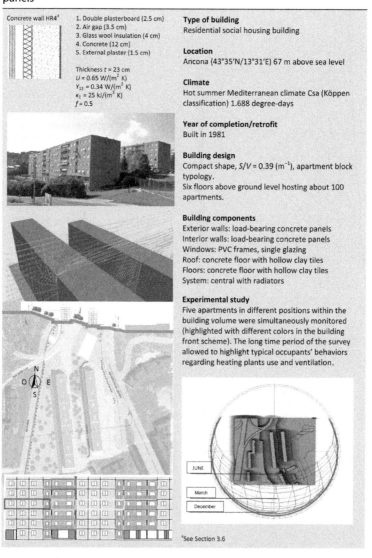

Concrete wall HR4[a]

1. Double plasterboard (2.5 cm)
2. Air gap (3.5 cm)
3. Glass wool insulation (4 cm)
4. Concrete (12 cm)
5. External plaster (1.5 cm)

Thickness t = 23 cm
U = 0.65 W/(m² K)
Y_{12} = 0.34 W/(m² K)
κ_1 = 25 kJ/(m² K)
f = 0.5

Type of building
Residential social housing building

Location
Ancona (43°35′N/13°31′E) 67 m above sea level

Climate
Hot summer Mediterranean climate Csa (Köppen classification) 1.688 degree-days

Year of completion/retrofit
Built in 1981

Building design
Compact shape, S/V = 0.39 (m⁻¹), apartment block typology.
Six floors above ground level hosting about 100 apartments.

Building components
Exterior walls: load-bearing concrete panels
Interior walls: load-bearing concrete panels
Windows: PVC frames, single glazing
Roof: concrete floor with hollow clay tiles
Floors: concrete floor with hollow clay tiles
System: central with radiators

Experimental study
Five apartments in different positions within the building volume were simultaneously monitored (highlighted with different colors in the building front scheme). The long time period of the survey allowed to highlight typical occupants' behaviors regarding heating plants use and ventilation.

JUNE
March
December

[a]See Section 3.6

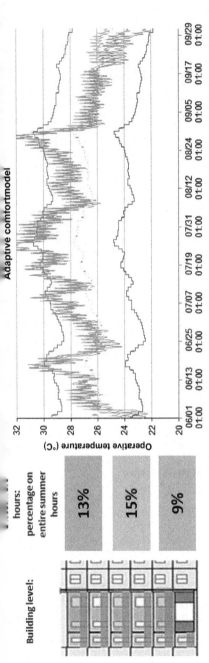

Figure A.10 Numerical simulations in summer. Percentage of hours in which the operative temperatures fall out of the comfort range at three building levels.

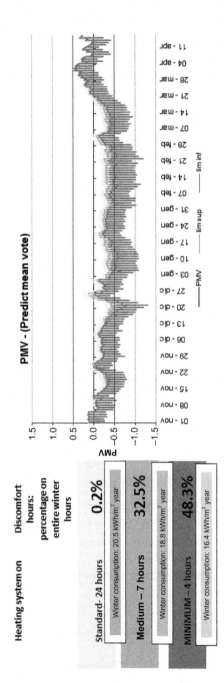

Figure A.11 Numerical simulations in winter. Percentage of hours in which the PMV values fall out of the range between −0.5 and 0.5.

Summer season performance

The detecting of the hours in which the operative temperatures fall without the comfort range defined by the adaptive method (see method in section 6.5.2) and the calculation of the relative percentage respect the hours of the entire summer period highlight that the first floor is not affected by overheating, showing a percentage under the acceptable value of 10% (Fig. A.10). The ultimate building level (sixth floor) is overheated for the adjacency of the roof slab that determines high incoming heat fluxes. However, the same roof provides cooling during the night, when the heat trapped inside could be expelled toward the sky. The worst condition (15% hours of discomfort) was recorded at mid-levels for the reduced heat loosing surfaces.

Winter season performance

The study of discomfort hours in an intermediate apartment on the building corner (*red rectangle* in the above front scheme) shows that the scarce use of the heating plant has a great incidence on indoor comfort levels (Fig. A.11). The extremely reduced use of the plants does not correspond to a noticeable reduction of the winter consumptions.

During the survey, four of the five monitored apartments of the social housing building were heated from users for a time shorter than 7 hours. Two of them for less than 4 hours.

As a consequence retrofit interventions on such buildings, through the introduction of additional insulation layers, may cause summer overheating phenomena (especially in densely occupied apartments) or be disadvantageous from global costs evaluations for the limited use of the plants.

B.6 Residential House, Villa Potenza, Macerata Italy

Table A.11 regards the simultaneous monitoring of two multistory buildings. They have identical features but only one of them was retrofitted in 1990 with an external insulation layer.

Summer season performance

The comparison between not insulated building and insulated one in typical summer days (Fig. A.12) shows that the external insulation is able to reduce the internal surface temperature values of about 3−4°C.

Winter season performance

The winter comparison reveals a beneficial increase of internal surface temperature for the retrofitted building up to 5°C (Fig. A.13).

Table A.11 Two multistory buildings with hollow cavity walls, one not insulated and one insulated

Unfilled brick cavity wall
S1[a]

1. Internal plaster (1.5 cm)
2. Perforated brick (8 cm)
3. Cement plaster (1.5 cm)
4. Air gap (9 cm)
5. Perforated brick (12 cm)

Thickness t = 32 cm
U = 1.2 W/(m² K)
Y_{12} = 0.5 W/(m² K)
κ_1 = 50 kJ/(m² K)
f = 0.41

Unfilled brick cavity wall
with existing external
insulation (retrofit 1990)

HR1[a]

1–5 as above
7. EPS insulation (5 cm)
8. Plaster (1.5 cm)

Thickness t = 40 cm
U = 0.4 W/(m² K)
Y_{12} = 0.11 W/(m² K)
κ_1 = 44 kJ/(m² K)
f = 0.26

Type of building
Two Residential buildings, built together with identical features.

Location
Villa Potenza (43°19′N/13°25′E) 97 m above sea level

Climate
Hot summer Mediterranean climate Csa (Köppen classification) 2.005 degree-days

Year of completion/retrofit
Built in 1974. Only one of the two volumes was retrofitted in 1990. So the external wall was changed from S1 to HR1.

Building design
Compact shape, S/V = 0.5 (m⁻¹), multistory building typology. Four floors above ground level; the ground floor hosts unheated garages.

Building components
Exterior walls: cavity walls (not insulated/insulated)
Interior walls: plastered perforated bricks 10 cm
Windows: PVC frames, single glazing
Roof: pitched roof with brick-concrete slab
Floors: reinforced brick-concrete slab
System: central with radiators

Experimental study
Two apartments (highlighted in *red* in the figure) in the same exposure within the two buildings (not insulated and insulated one) were simultaneously monitored to deepen the effect of the external insulation on existing cavity walls.

1. Flux meter
2. Surface temperature PT100
3. Hydrothermal probe
4. Sun radiation
5. Air temperature at different heights
6. Combined air temperature and RH sensor
7. Black globe temperature probe
8. Anemometer

[a]See Sections 3.3 and 3.5

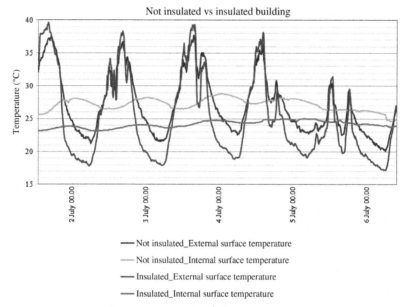

Figure A.12 Results of the summer monitoring campaign. Comparison between building with walls S1 and HR1.

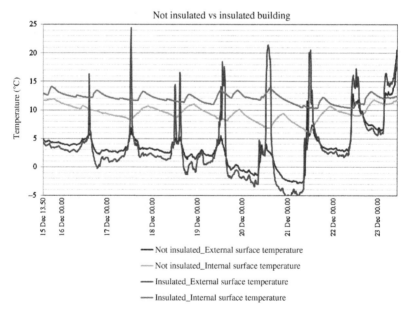

Figure A.13 Results of the winter monitoring campaign. Comparison between building with walls S1 and HR1.

B.7 School classrooms, Loreto, Ancona, Italy

Table A.12 regards a monitoring of a school building to deepen the effect of the presence of high mass in the envelope internal side, especially in densely occupied environments.

Table A.12 Compact volume with uninsulated cavity walls

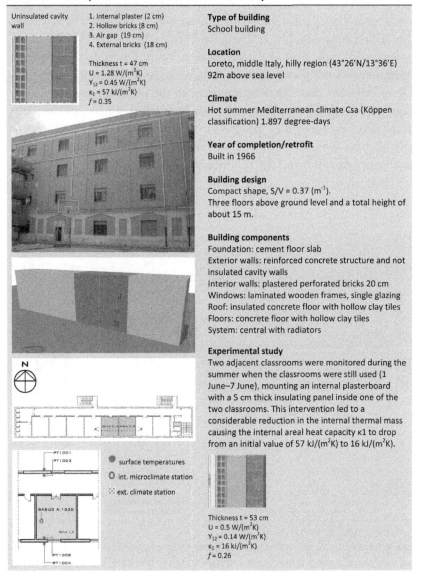

Uninsulated cavity wall

1. Internal plaster (2 cm)
2. Hollow bricks (8 cm)
3. Air gap (19 cm)
4. External bricks (18 cm)

Thickness t = 47 cm
U = 1.28 W/(m²K)
Y_{12} = 0.45 W/(m²K)
K_1 = 57 kJ/(m²K)
f = 0.35

Type of building
School building

Location
Loreto, middle Italy, hilly region (43°26'N/13°36'E)
92m above sea level

Climate
Hot summer Mediterranean climate Csa (Köppen classification) 1.897 degree-days

Year of completion/retrofit
Built in 1966

Building design
Compact shape, S/V = 0.37 (m⁻¹).
Three floors above ground level and a total height of about 15 m.

Building components
Foundation: cement floor slab
Exterior walls: reinforced concrete structure and not insulated cavity walls
Interior walls: plastered perforated bricks 20 cm
Windows: laminated wooden frames, single glazing
Roof: insulated concrete floor with hollow clay tiles
Floors: concrete floor with hollow clay tiles
System: central with radiators

Experimental study
Two adjacent classrooms were monitored during the summer when the classrooms were still used (1 June–7 June), mounting an internal plasterboard with a 5 cm thick insulating panel inside one of the two classrooms. This intervention led to a considerable reduction in the internal thermal mass causing the internal areal heat capacity κ1 to drop from an initial value of 57 kJ/(m²K) to 16 kJ/(m²K).

N

• surface temperatures
O int. microclimate station
⊠ ext. climate station

Thickness t = 53 cm
U = 0.5 W/(m²K)
Y_{12} = 0.14 W/(m²K)
K_1 = 16 kJ/(m²K)
f = 0.26

Summer performance: comparison between two levels of κ_1

The summer monitoring on the two classrooms (with and without internal insulation layer) allowed demonstrating the incidence of the internal mass.

The graph in Fig. A.14 shows how the trends in internal air temperature of the two classrooms are inverted as soon as the pupils occupy the room (*yellow hatch* (light gray in print versions)). Indeed the wall with high internal mass keeps the temperature lower when the room is occupied accumulating heat which is released when the pupils leave the classroom. This is clearly visible from the curve of heat fluxes. Given that the external temperatures are not elevated the flows are almost always outgoing. So for this graph the flux directed from the room to the external envelope is considered with positive sign. The wall with high internal mass on the inner surface (*black line*) presents elevated storing attitude when the room is occupied with flow toward outdoors until about 3 p.m. After this time the heat flow changes direction and the wall releases the accumulated heat inward.

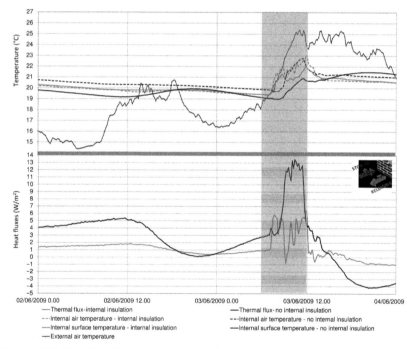

Figure A.14 Results of the simultaneous monitoring of two school classrooms with envelopes characterized by different internal mass (in one case it was mounted an internal insulation layer).

C NEW ENVELOPES

C.1 Residential House, Chiaravalle, Ancona, Italy

Table A.13 regards the simultaneous monitoring of two recently built volumes, one with a lightweight timber envelope, the other with massive brick walls.

Table A.13 Two very similar and adiacent isolated buildings with respectively platform framed envelope and semisolid brick walls

Platform frame wooden envelope

1. Internal plaster (1.5 cm)
2. Perforated bricks (8 cm)
3. OSB panel (1.8 cm)
4. Air gap (11 cm)
5. Rock wool insulation (5 cm)
6. OSB panel (1.8 cm)
7. EPS insulation (10 cm)
8. External plaster (1.5 cm)

Thickness t = 40.6 cm
U = 0.22 W/(m² K)
Y_{12} = 0.05 W/(m² K)
κ_1 = 50 kJ/(m² K)
f = 0.21

Type of building
Two-families house

Location
Falconara, middle Italy, eastern coast
(43°35'N/13°30'E) 0–200 m above sea level

Climate
Hot summer Mediterranean climate Csa (Köppen classification) 1.823 degree-days

Year of completion
Built in 2009

Building design
Compact shape, S/V = 0.88 (m⁻¹).
One floor above ground level with attic floor

Building components
Foundation: cement floor slab
Exterior walls: platform frame system with the addition of an internal massive layer
Interior walls: insulation and plasterboards
Windows: PVC frames, double glazing
Roof: wooden plank, EPS insulation(16 cm), OSB panel and clay tiles
System: underfloor heating

Experimental study
A monitoring campaign regarded the simultaneous survey of the present case study and another building adjacent to it, with the following envelope:

Semisolid bricks and external insulation layer

Thickness t = 40 cm
U = 0.27 W/(m² K)
Y_{12} = 0.02 W/(m² K)
κ_1 = 43 kJ/(m² K)
f = 0.07

The two buildings present the same geometry and features except for the technique adopted for the external envelope.

Blower door test

Realization phases

The images in Fig. A.15 refer to the platform framed building realization phases. The assembly is done in short time periods but careful attention should be paid on connections and waterproofing.

Figure A.15 Platform framed envelope: realization phases.

Summer performance: comparison with a capacity envelope

The comparison between two buildings having the same geometry and internal distribution but characterized by two different envelope techniques, wood and masonry, allowed highlighting that the adoption of inner massive finishing materials could enhance the performance of a lightweight envelope. The graph in Fig. A.16 shows that unexpectedly the platform framed envelope (Wood) behaves better than the masonry ones with lower operative temperatures. However, the occupants' habits had great incidence as stressed by the graphs in Fig. A.17 that shows a very different radiation recorded inside the two buildings (almost absent in the wooden one) due to a different use of the sun shadings and to different occupancy profiles.

So to make a more reliable comparison, a numerical simulation adopting the same internal gains is needed.

The simulations confirmed that the wooden solution with internal massive lining has a good summer behavior, similar to masonry building (Fig. A.18). However, adopting the same internal gains the latter has higher comfort levels, for its more favorable dynamic parameters, combining a high κ_1 value with a low f value.

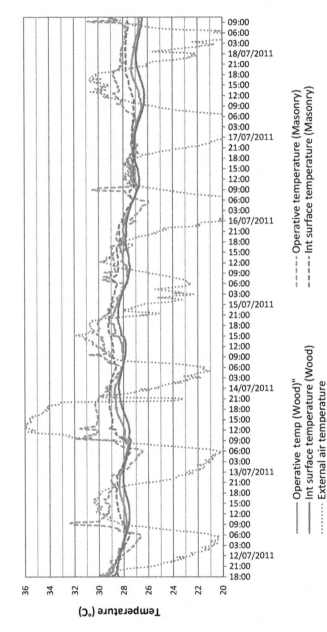

Figure A.16 Results of the simultaneous monitoring of two buildings characterized by either a masonry envelope or a platform framed envelope.

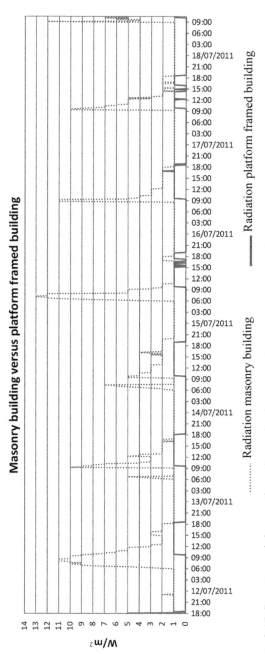

Figure A.17 Experimental data on the radiation levels measured inside the two buildings.

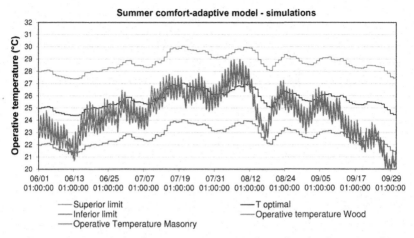

Figure A.18 Numerical simulations. Comparison of comfort levels within the two buildings with the same internal gains.

C.2 Residential house, Barbara, Ancona, Italy

Table A.14 regards the monitoring of a building entirely realized with the wood-cement technique.

Realization phases

The images in Fig. A.19 refer to the building realization phases. The wood-cement technique is very appreciated in recent years for its high seismic resistance and the elevated thermal and fire protection performance.

Summer performance

The monitoring at the beginning of the summer period (Fig. A.20) highlighted a comfortable indoor environment even if the external temperatures in some days reached high values.

Table A.14 Single family house with wood-cement envelope

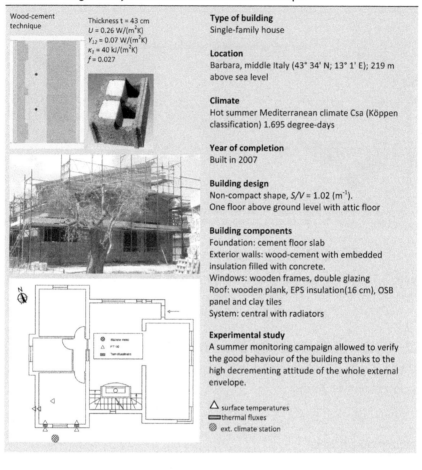

Wood-cement technique	Thickness t = 43 cm U = 0.26 W/(m²K) Y_{12} = 0.07 W/(m²K) κ_1 = 40 kJ/(m²K) f = 0.027

Type of building
Single-family house

Location
Barbara, middle Italy (43° 34' N; 13° 1' E); 219 m above sea level

Climate
Hot summer Mediterranean climate Csa (Köppen classification) 1.695 degree-days

Year of completion
Built in 2007

Building design
Non-compact shape, S/V = 1.02 (m⁻¹).
One floor above ground level with attic floor

Building components
Foundation: cement floor slab
Exterior walls: wood-cement with embedded insulation filled with concrete.
Windows: wooden frames, double glazing
Roof: wooden plank, EPS insulation(16 cm), OSB panel and clay tiles
System: central with radiators

Experimental study
A summer monitoring campaign allowed to verify the good behaviour of the building thanks to the high decrementing attitude of the whole external envelope.

△ surface temperatures
▭ thermal fluxes
◉ ext. climate station

Figure A.19 Wood-cement envelope: realization phases.

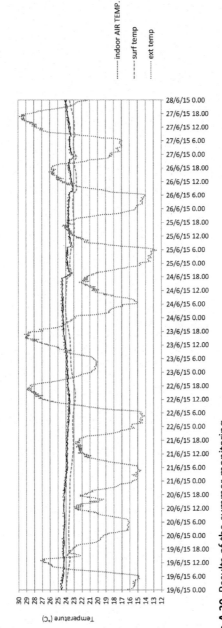

Figure A.20 Results of the summer monitoring.

C.3 Residential house, Porto Sant'Elpidio, Fermo, Italy

Table A.15 regards the monitoring of a superinsulated passive house building.

Table A.15 Single family passive house with a Platform framed envelope

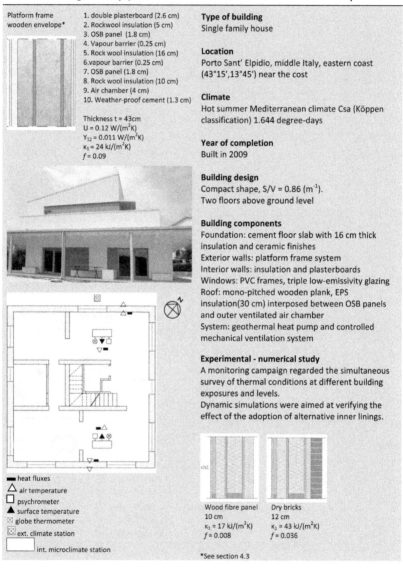

Platform frame wooden envelope*

1. double plasterboard (2.6 cm)
2. Rockwool insulation (5 cm)
3. OSB panel (1.8 cm)
4. Vapour barrier (0.25 cm)
5. Rock wool insulation (16 cm)
6. vapour barrier (0.25 cm)
7. OSB panel (1.8 cm)
8. Rock wool insulation (10 cm)
9. Air chamber (4 cm)
10. Weather-proof cement (1.3 cm)

Thickness t = 43cm
$U = 0.12$ W/(m^{2}K)
$Y_{12} = 0.011$ W/(m^{2}K)
$\kappa_1 = 24$ kJ/(m^{2}K)
$f = 0.09$

Type of building
Single family house

Location
Porto Sant' Elpidio, middle Italy, eastern coast
(43°15′,13°45′) near the cost

Climate
Hot summer Mediterranean climate Csa (Köppen classification) 1.644 degree-days

Year of completion
Built in 2009

Building design
Compact shape, S/V = 0.86 (m^{-1}).
Two floors above ground level

Building components
Foundation: cement floor slab with 16 cm thick insulation and ceramic finishes
Exterior walls: platform frame system
Interior walls: insulation and plasterboards
Windows: PVC frames, triple low-emissivity glazing
Roof: mono-pitched wooden plank, EPS insulation(30 cm) interposed between OSB panels and outer ventilated air chamber
System: geothermal heat pump and controlled mechanical ventilation system

Experimental - numerical study
A monitoring campaign regarded the simultaneous survey of thermal conditions at different building exposures and levels.
Dynamic simulations were aimed at verifying the effect of the adoption of alternative inner linings.

■ heat fluxes
△ air temperature
□ psychrometer
▲ surface temperature
⊠ globe thermometer
⊠ ext. climate station
☐ int. microclimate station

Wood fibre panel
10 cm
$\kappa_1 = 17$ kJ/(m^{2}K)
$f = 0.008$

Dry bricks
12 cm
$\kappa_1 = 43$ kJ/(m^{2}K)
$f = 0.036$

*See section 4.3

Source: F. Stazi, E. Tomassoni, C. Di Perna, Super-insulated wooden envelopes in Mediterranean climate: summer overheating, thermal comfort optimization, environmental impact on an Italian case study, Energy Build. 138 (March 1, 2017) 716–732, ISSN 0378-7788.

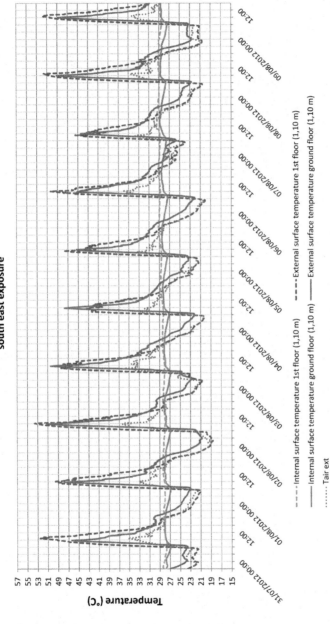

Figure A.21 Results of the summer monitoring at the two building levels.

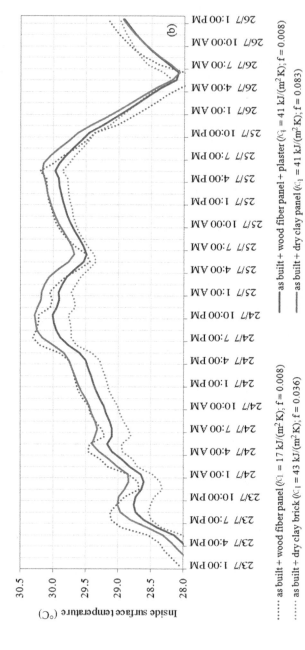

Figure A.22 Numerical simulations. Inside surface temperatures by adopting different internal linings. The envelope characterized by a high κ_1 value combined with low (but not excessively low) value of decrement factor achieves the best performance (see explanation in Chapter 2: The Envelope: A Complex and Dynamic Problem). *From F. Stazi, E. Tomassoni, C. Di Perna, Super-insulated wooden envelopes in Mediterranean climate: summer overheating, thermal comfort optimization, environmental impact on an Italian case study, Energy Build. 138 (March 1, 2017) 716–732, ISSN 0378-7788.*

····· as built + wood fiber panel (κ_1 = 17 kJ/(m²K); f = 0.008)
····· as built + dry clay brick (κ_1 = 43 kJ/(m²K); f = 0.036)

—— as built + wood fiber panel + plaster (κ_1 = 41 kJ/(m²K); f = 0.008)
····· as built + dry clay panel (κ_1 = 41 kJ/(m²K); f = 0.083)

Summer performance

The experimental study of the summer behavior of the building at the two levels highlighted high indoor temperatures at both levels (Fig. A.21). The ground floor room is affected by higher fluctuations for the presence of heat fluxes toward the ground.

The simulations revealed that the envelope could be optimized by adopting internal linings with high κ_1 and low attenuating attitude. As an example the graph in Fig. A.22 reports the comparison between three types of internal linings: the best behavior is shown by the as built wall combined with an internal dry clay brick, with lower inside (and operative) surface temperatures. Indeed this solution has the most favorable dynamic parameters combination.

D PASSIVE ENVELOPES

D.1 Office building with ventilated clay walls, Ancona, Italy

Table A.16 regards a monitoring of ventilated facades with an open joints clay cladding.

Winter performance

In winter (Fig. A.23) the ventilated wall shows an effective ventilation, with air speeds around 0.6 m/s in the central hours of the day, decreasing down to 0.2 m/s in the night.

Summer performance

In summer (Fig. A.24) the wall shows similar conditions of the winter period, with slightly higher air speeds, around 0.8 m/s, concentrated in the central hours of the day.

Table A.16 Office building with ventilated clay walls, open joints

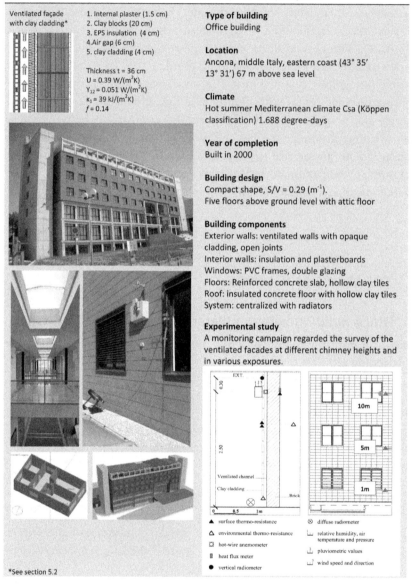

Ventilated façade with clay cladding*

1. Internal plaster (1.5 cm)
2. Clay blocks (20 cm)
3. EPS insulation (4 cm)
4. Air gap (6 cm)
5. clay cladding (4 cm)

Thickness t = 36 cm
U = 0.39 W/(m²K)
Y_{12} = 0.051 W/(m²K)
κ_1 = 39 kJ/(m²K)
f = 0.14

Type of building
Office building

Location
Ancona, middle Italy, eastern coast (43° 35′ 13° 31′) 67 m above sea level

Climate
Hot summer Mediterranean climate Csa (Köppen classification) 1.688 degree-days

Year of completion
Built in 2000

Building design
Compact shape, S/V = 0.29 (m⁻¹).
Five floors above ground level with attic floor

Building components
Exterior walls: ventilated walls with opaque cladding, open joints
Interior walls: insulation and plasterboards
Windows: PVC frames, double glazing
Floors: Reinforced concrete slab, hollow clay tiles
Roof: insulated concrete floor with hollow clay tiles
System: centralized with radiators

Experimental study
A monitoring campaign regarded the survey of the ventilated facades at different chimney heights and in various exposures.

▲ surface thermo-resistance
△ environmental thermo-resistance
◻ hot-wire anemometer
▌ heat flux meter
● vertical radiometer

⊗ diffuse radiometer
⊔ relative humidity, air temperature and pressure
⊔ pluviometric values
⊔ wind speed and direction

*See section 5.2

Source: From F. Stazi, F. Tomassoni, A. Vegliò, C. Di Perna, Experimental evaluation of ventilated walls with an external clay cladding, Renewable Energy 36 (12) (December 2011) 3373–3385, ISSN 0960-1481.

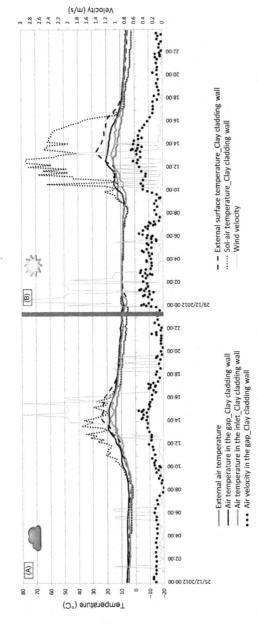

Figure A.23 Results of the winter monitoring in (A) cloudy and (B) sunny days.

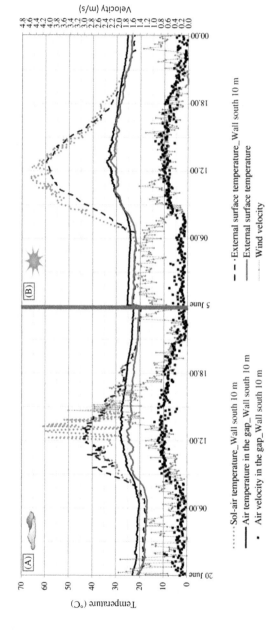

Figure A.24 Results of the summer monitoring in (A) cloudy and (B) sunny days.

D.2 Office and school building with ventilated titanium walls, Ancona, Italy

See Table A.17 regards a monitoring of ventilated facades with a closed joint zinc titanium cladding.

Table A.17 Office building with ventilated zinc titanium walls, closed joints

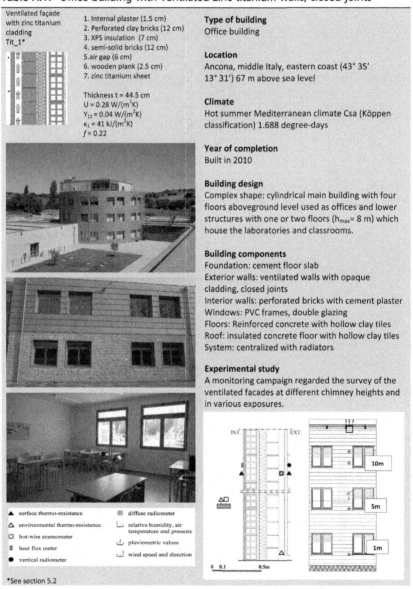

Ventilated façade with zinc titanium cladding Tit_1*

1. Internal plaster (1.5 cm)
2. Perforated clay bricks (12 cm)
3. XPS insulation (7 cm)
4. semi-solid bricks (12 cm)
5. air gap (6 cm)
6. wooden plank (2.5 cm)
7. zinc titanium sheet

Thickness t = 44.5 cm
U = 0.28 W/(m²K)
Y_{12} = 0.04 W/(m²K)
κ_1 = 41 kJ/(m²K)
f = 0.22

Type of building
Office building

Location
Ancona, middle Italy, eastern coast (43° 35′ 13° 31′) 67 m above sea level

Climate
Hot summer Mediterranean climate Csa (Köppen classification) 1.688 degree-days

Year of completion
Built in 2010

Building design
Complex shape: cylindrical main building with four floors aboveground level used as offices and lower structures with one or two floors (h_{max}= 8 m) which house the laboratories and classrooms.

Building components
Foundation: cement floor slab
Exterior walls: ventilated walls with opaque cladding, closed joints
Interior walls: perforated bricks with cement plaster
Windows: PVC frames, double glazing
Floors: Reinforced concrete with hollow clay tiles
Roof: insulated concrete floor with hollow clay tiles
System: centralized with radiators

Experimental study
A monitoring campaign regarded the survey of the ventilated facades at different chimney heights and in various exposures.

▲ surface thermo-resistance
△ environmental thermo-resistance
◻ hot-wire anemometer
▌ heat flux meter
● vertical radiometer

⊗ diffuse radiometer
⨆ relative humidity, air temperature and pressure
⨆ pluviometric values
⨆ wind speed and direction

*See section 5.2

Source: From F. Stazi, A. Vegliò, C. Di Perna, Experimental assessment of a zinc–titanium ventilated façade in a Mediterranean climate, Energy Build. 69 (February 2014) 525–534, ISSN 0378-7788.

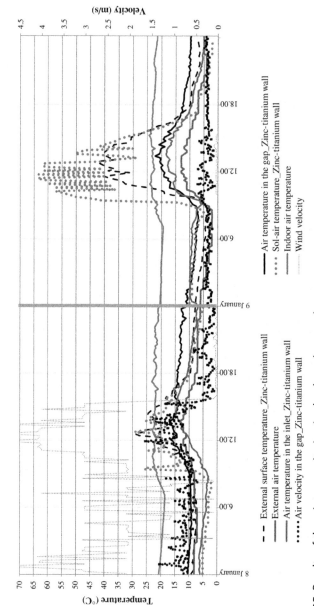

Figure A.25 Results of the winter monitoring in cloudy and sunny days.

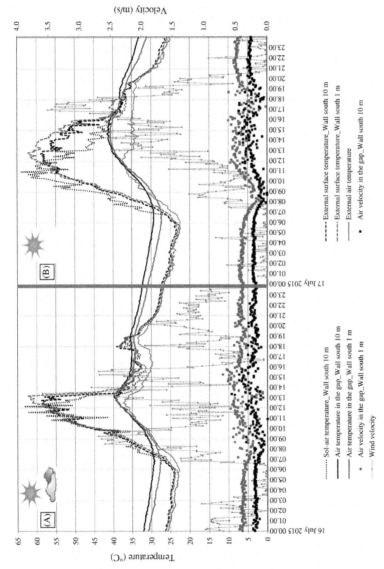

Figure A.26 Results of the summer monitoring in (A) cloudy and (B) sunny days.

Winter performance

In a sunny winter day (Fig. A.25) the wall shows a moderate ventilation rate, with air speeds around 0.3–0.4 m/s throughout the day and night. In a cloudy day with high wind (from north) the ventilation is even more effective.

Summer performance

In summer (Fig. A.26) the walls show similar conditions of the winter period, with slightly higher air speed values, between 0.4 and 0.8 m/s, throughout the day.

The m wall has similar air temperatures and velocity in the gap on the two days, regardless of the presence or absence of clouds.

D.3 Residential building with passive solar walls, Ancona, Italy

Table A.18 regards the monitoring of a passive house with solar systems.

Winter performance

The winter monitoring in unshaded and unvented conditions (Fig. A.27) demonstrated the presence of a very comfortable indoor environment. The time lag of 12 hours determines that the maximum values reached on the external surface between 1 p.m. and 3 p.m. cause an increase of the internal surface temperature at night, aiding the heating system in maintaining indoor comfortable temperatures.

Summer performance

The summer monitoring in shaded and vented conditions (Fig. A.28) showed that the surface temperature of the concrete wall on the external side reaches the maximum value of 27°C. The internal air temperatures are stable around 24°C. During the day the wall stores the heat having a surface temperature value lower than the ambient temperatures. During the night it has instead higher values.

The thermographic survey (Fig. A.29) highlighted the presence of a vertical thermal gradient with the highest values recorded at wall mid-height in correspondence of the air vents.

Table A.18 The monitored apartment with Trombe walls on the southern exposure

Trombe wall*	
	1. Internal plaster (1 cm)
	2. Concrete wall painted black on the outer side (40 cm)
	3. air gap (10 cm)
	4. single glazing (0.4 cm)
	Thickness t = 51 cm
	U = 1.60 W/(m²K) unshaded
	U = 1.88 W/(m²K) shaded + vented
	Y_{12} = 0.13 W/(m²K)
	κ_1 = 75 kJ/(m²K)
	f = 0.08

Type of building
Residential building

Location
Ancona, middle Italy, eastern coast (43° 35′ 13° 31′) 67 m above sea level

Climate
Hot summer Mediterranean climate Csa (Köppen classification) 1.688 degree-days

Year of completion
Built in 1983

Building design
Compact shape: 0.5 (m⁻¹)
Three floors above ground level

Building components
Foundation: cement floor slab
Exterior walls: passive solar system on southern side
Windows: PVC frames, single glazing
Floors: Reinforced concrete slab, hollow clay tiles
Roof: insulated concrete floor with hollow clay tiles
System: centralized with radiators

Experimental study
A monitoring campaign regarded the survey of the Trombe walls performance in different seasons, occupancy conditions, and under different ventilation regimes. Moreover the effect of the use of solar shadings in summer was evaluated.

*See section 5.3

Source: From F. Stazi, A. Mastrucci, C. Di Perna, The behaviour of solar walls in residential buildings with different insulation levels: an experimental and numerical study, Energy Build. 47 (April 2012) 217–229, ISSN 0378-7788.

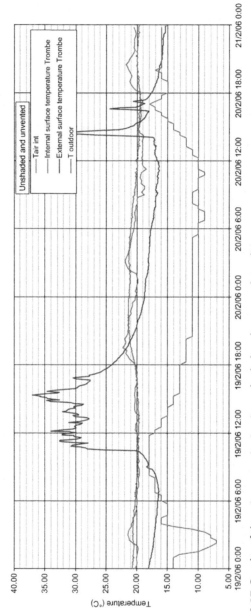

Figure A.27 Results of the winter monitoring in unshaded and unvented condition.

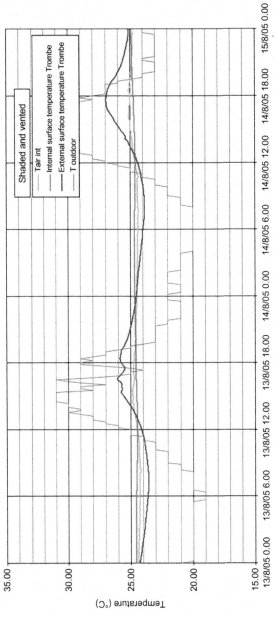

Figure A.28 Results of the summer monitoring in shaded and vented condition.

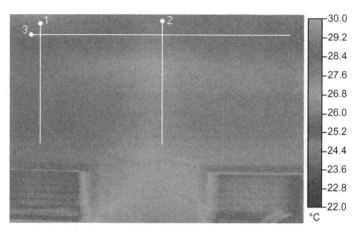

	Line 1	Line 2	Line 3
Emissivity	0.90	0.90	0.90
Avg temp (°C)	28.2	27.3	28.3
Min temp (°C)	27.8	26.9	27.6
Max temp (°C)	29.1	27.9	28.9
Delta ref. (°C)			

Figure A.29 Result of the thermographic survey. *From F. Stazi, A. Mastrucci, C. Di Perna, Trombe wall management in summer conditions: an experimental study, Solar Energy 86 (9) (September 2012) 2839–2851, ISSN 0038-092X.*

Details on Numerical Methods

I INTRODUCTION

The following equations and descriptions regard the detailing of methods adopted for the calculation of the dynamic thermal parameters according to EN13786:2007 and for the thermal simulations in dynamic regime with the software EnergyPlus.

B.1 CALCULATION OF DYNAMIC THERMAL PARAMETERS

The calculation method described hereafter is based on EN13786. It considers the thermal conduction in multilayered building components composed of parallel and plane homogeneous layers. The assumption is that the variations of temperature and heat flows around their long-term average could be described by a sine function of time. Thermal bridges usually present in such components do not affect the dynamic thermal performance in a significant manner, so are neglected.

The physical quantities of one side of the component are related to those on its other side. This calculation allows obtaining the heat storage capacity of a given component.

Step 0

Define input time period

The values of the dynamic thermal characteristics depend on the selected period of thermal variation T. So a period of thermal variation T should be chosen. A period of 86,400 seconds corresponds to daily meteorological variations.

Step 1. Thermal properties of each "i" layer

Layers materials and thickness d (m) for each layer.

The selection of materials makes it is possible to identify the following thermal characteristics:

Thermal conductivity, λ (W/(m K)).

Specific heat capacity, c (J/(kg K)).

Density, ρ (kg/m^3).

Periodic penetration depth and thermal resistances.

It is now possible to calculate for each layer the following parameters:

1. $\delta_i = \sqrt{\lambda T / \pi \rho c}$ Periodic penetration depth of a heat wave in a material (m).

   It represents the depth at which the amplitude of temperature variations are reduced by the factor "e." This factor is the base of natural logarithms ($e = 2.718$, etc.). So it describes the decay of thermal wave inside of a material.

2. $\xi_i = d/\delta$ *Dimensionless ratio of the layer thickness to the penetration depth (−).*

3. $R_i = d/\lambda$ *Thermal resistance (m^2 K/W).*

4. R_{air} *Air layer thermal resistance (m^2 K/W).*

   In the case of the presence of an air layer (e.g., within a cavity wall), its thermal resistance should be calculated according to ISO 6946, with different values based on flow direction.

5. $R_{s1,2}$ *Thermal resistance of surface boundary layer ((m^2 K)/W).*

   R_{s1} and R_{s2} are the surface resistances of, respectively, the internal (1) and external (2) boundary layer, including convection and radiation, calculated according to ISO 6946.

Step 2. Steady-state thermal parameters for the building component

1. R_t *Total thermal resistance ((m^2 K)/W).*

   It is the sum of the resistance of all the layers.

$$R_t = R_{s1} + \sum R_i + R_{air} + R_{s2}$$

2. $U_0 = 1/R_t$ *Thermal transmittance ((W/(m^2 K)).*

   It is calculated ignoring the thermal bridges.

3. $\kappa_m = \sum \rho_i d_i c_i$ *Long-term (steady state) capacity (kJ/(m^2 K)).*

   Note that adopting a specific heat capacity in Joule c (J/kg K), the value should be divided by 1000.

Step 3. Heat transfer matrix of each homogeneous layer

The complex amplitudes (see note$\star$) of temperature $\hat{\theta}_2$ and heat flow rate $\hat{q}_2$ on one side of each layer "i," could be related with the complex amplitudes of temperature $\hat{\theta}_1$ and heat flow rate $\hat{q}_1$ on the other side through the heat transfer matrix:

$$\begin{pmatrix} \hat{\theta}_2 \\ \hat{q}_2 \end{pmatrix} = \begin{pmatrix} Z_{11} & Z_{12} \\ Z_{21} & Z_{22} \end{pmatrix} \begin{pmatrix} \hat{\theta}_1 \\ \hat{q}_1 \end{pmatrix} \tag{B.1}$$

HEAT TRANSFER MATRIX of i layer: $Z_i (i= 1...N)$

The transfer matrix elements are calculated as follows:

REAL PART (a)   IMAGINARY PART (b)

$$Z_{11} = Z_{22} = \cosh(\xi)\cos(\xi) + j\sinh(\xi)\sin(\xi) \tag{B.2}$$

$$Z_{12} = -\frac{\delta}{2\lambda}\{\sinh(\xi)\cos(\xi) + \cosh(\xi)\sin(\xi) + j[\cosh(\xi)\sin(\xi) - \sinh(\xi)\cos(\xi)]\} \tag{B.3}$$

$$Z_{21} = -\frac{\lambda}{\delta}\{\sinh(\xi)\cos(\xi) - \cosh(\xi)\sin(\xi) + j[\sinh(\xi)\cos(\xi) + \cosh(\xi)\sin(\xi)]\} \tag{B.4}$$

The calculations with complex number are replaced by conventional matrices calculation.

Each of these elements determines a matrix of order 2 (see the note$\star$ on complex numbers). So the final heat transfer matrix Z_i for each layer "i" (formula B.1) will be of order 4.

If the layer is an air cavity, then the specific heat capacity is neglected and the corresponding matrix is the following:

$$(Z_{air}) = \begin{pmatrix} 1 & -R_{air} \\ 0 & 1 \end{pmatrix} \tag{B.5}$$

Step 4. From layers to the multilayer component: product of matrices

$$Z = Z_N Z_{N-1}......Z_3 Z_2 Z_1 \tag{B.6}$$

Z is the HEAT TRANSFER MATRIX of the entire component. It is calculated as the product of the heat transfer matrices of the various layers. The calculation should begin from the innermost layer (layer 1). Then it is possible to obtain:

$$Z_{ee} = Z_{s2} ZZ_{s1}; \qquad (B.7)$$

Z_{ee} *is the heat transfer matrix from environment to environment through the building component.* It includes in the calculation the heat transfer matrix of the boundary layers:

$$Z_{s1} = \begin{pmatrix} 1 & -R_{s1} \\ 0 & 1 \end{pmatrix} \text{ (Heat transfer matrix of the internal boundary layer)}$$

$$(B.8)$$

$$Z_{s2} = \begin{pmatrix} 1 & -R_{s2} \\ 0 & 1 \end{pmatrix} \text{ (Heat transfer matrix of the external boundary layer)}$$

$$(B.9)$$

Step 5. Dynamic thermal characteristics calculation

The matrices calculation allows obtaining the dynamic thermal characteristics. The passage from matrices to real numbers is made through the modulus and argument (see note on complex numbers).

The thermal admittances are:

$$Y_{11} = -\frac{Z_{11}}{Z_{12}}; \quad \Delta t_{Y_{11}} = \frac{T}{2\pi} \, arg(Y_{11}) \qquad (B.10)$$

$$Y_{22} = -\frac{Z_{22}}{Z_{12}}; \quad \Delta t_{Y_{22}} = \frac{T}{2\pi} \, arg(Y_{22}) \qquad (B.11)$$

The areal heat capacities are:

$$\kappa_1 = \frac{T}{2\pi} \left| \frac{Z_{11} - 1}{Z_{12}} \right| \text{ internal side} \qquad (B.12)$$

and

$$\kappa_2 = \frac{T}{2\pi} \left| \frac{Z_{22} - 1}{Z_{12}} \right| \text{ external side} \qquad (B.13)$$

The periodic thermal transmittance is given by:

$$Y_{12} = -\frac{1}{Z_{12}} \qquad (B.14)$$

The decrement factor is:

$$f = \frac{|Y_{12}|}{U_0} \tag{B.15}$$

Note*. Complex numbers

The heat transfer within a building component is computed with complex numbers.

The complex numbers can be expressed in algebraic form through the following relation:

$$z = a + ib \tag{B.16}$$

where a is the real part and b is the imaginary part of the complex number; i (or sometimes j) is the imaginary unit, satisfying the equation $i^2 = -1$ (Fig. B.1).

A complex number can be also represented in matrix notation:

$$z = \begin{pmatrix} a & b \\ -b & a \end{pmatrix} \tag{B.17}$$

So a complex number coincides with a real matrix of order 2.

The modulus is calculated with the following relation:

$$|z| = \sqrt{a^2 + b^2} \tag{B.18}$$

The argument, useful for time shift calculations (see previous points B.10 and B.11), can be obtained with arctan function adopting different formulas according to the values assumed by a and b.

For example, for $a > 0, b > 0$ the formula to calculate the time shift of admittance is $\arg(z) = \arctan(b/a)$ and the formula to calculate the time shift of periodic thermal transmittance is $\arg(z) = \arctan(b/a) - 2\pi$.

Table B.1 reports an example of calculation for a wall studied within the book.

Figure B.1 Representation of complex numbers.

Table B.1 Example of calculation of dynamic parameters

EXAMPLE. Envelope C2$_{ext}$. Example of calculation.

Step 0. The following examples refer to a time period of 24 h (84,600 s).

Steps 1–2. Thermal properties of each "i" layer—steady state thermal parameters for the building component

Layers (from internal to external side)	Layer thickness	Thermal conduct.	Specific heat	Material density	Periodic penetration depth	$d/(\delta)$	Thermal resistance
	d (m)	λ (W/m K)	c (J/Kg K)	ρ (kg/m^3)	δ (m)	ξ (–)	R (m^2 K/W)
Internal boundary							0.130 (R_{s1})
1 Internal plaster	0.015	0.900	1000	1900	0.114	0.131	0.017
2 Solid bricks	0.420	0.780	940	1500	0.123	3.405	0.538
3 plaster	0.015	0.900	1000	1900	0.114	0.131	0.017
4 EPS insulation	0.120	0.036	1480	35	0.138	0.868	3.333
5 External plaster	0.005	0.700	1000	1000	0.139	0.036	0.007
External boundary							0.04 (R_{s2})

Steady-state thermal resistance	R_t ((m^{2}K/)W)	4.082
Steady-state thermal transmittance	U ((m^{2}K)/W)	0.245
Long-term steady state capacity	κ_m (kJ/(m^{2}K))	660

Step 3. Heat transfer matrix of each homogeneous layer (in both directions)

Matrix	Element of matrix	Modulus	Time shift (in range—12–12 h)
Heat transfer matrix	Z_{11}	660.84	−5.81
	Z_{21}	395.60 W/(m^2 K)	10.04
	Z_{12}	142.64 m^2 K/W	4.77
	Z_{22}	85.39	−3.38
Inverse matrix	Z'_{11}	85.39	−3.38
	Z'_{21}	395.60 W/(m^2 K)	−1.96
	Z'_{12}	142.64 m^2 K/W	−7.23
	Z'_{22}	660.84	−5.81

Step 4. Calculation of the dynamic thermal characteristics

Property	Modulus	Time shift (h)
Y_{11}	4.63	1.41
Y_{22}	0.60	3.85
Y_{12}	0.007	16.77
κ_1	63.7	—
κ_2	8.2	—
f	0.029	

B.2 NUMERICAL SIMULATIONS OF THE THERMAL PERFORMANCE OF THE WHOLE BUILDING

A virtual model reproducing the *as built* solution was realized for each case study (Fig. B.2).

The energy analyses in dynamic regime were carried out with *EnergyPlus* software of the United States Department of Energy [1]. This section reports some useful information on building simulations.

Basics

EnergyPlus (version 8.5) is a simulation tool, in which the input and output data are provided as a "text-based" idf file. Heat balance, indoor temperature, and comfort condition prediction are obtained by means of a complex algorithm resolution procedure. The Conduction Transfer Function algorithm allows to evaluate the surface heat fluxes linearly relating the heat transfer to the current and previous temperature levels and heat fluxes, so providing the evaluation of the thermal storage phenomena. Since EnergyPlus is a simulation engine with no interface, Design Builder (version 4.7.0.027) [2] was used as graphical interface to create the model and reproduce the complex building geometry, materials, constructions, and schedules for all time-dependent values.

The ventilation system was simulated through EnergyPlus Airflow Network Tool [3,4]. An airflow network consists of a set of nodes connected by airflow elements (Figs. B.3 and B.4).

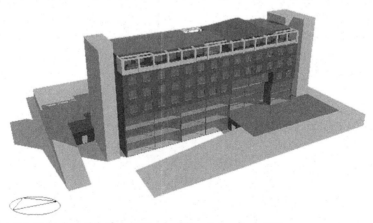

Figure B.2 Virtual model of an entire building (case study D1).

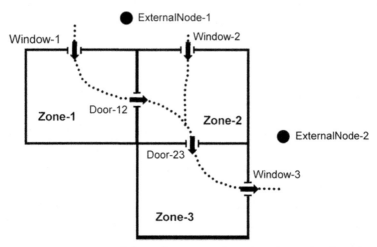

Figure B.3 Representation of the Airflow Network Tool: nodes and airflow elements taken by Ref. [3]. *From U.S. Department of Energy, DOE Energy Plus 7.0 Input Output Reference: The Encyclopedic Reference to Energy Plus Input and Output, U.S. Department of Energy, 2011.*

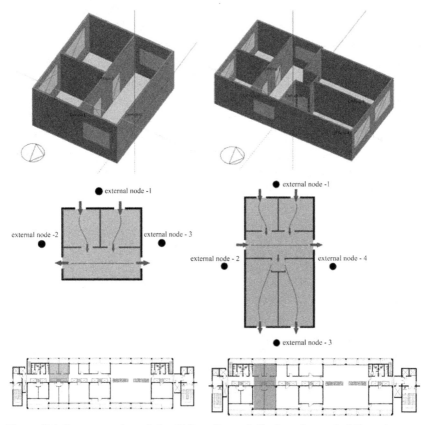

Figure B.4 Representation of the Airflow Network Tool: nodes and airflow elements adopting different thermal zones.

The nodes represent rooms, connection points in ductwork, or the ambient environment. The airflow elements correspond to discrete airflow passages such as doorways, construction cracks, ducts, and fans. The assumption is made that there is a simple, nonlinear relationship between the flow through a connection and the pressure difference across it. The problem therefore reduces to the calculation of the airflow through these connections with the internal nodes of the network representing certain unknown pressures. The pressure difference between two nodes and the external wind pressure is assumed to be governed by Bernoulli's equation. A solution is achieved by an iterative mass balance technique in which the unknown nodal air pressures are adjusted by applying mass conservation. Based on the found airflows the model then calculates node temperatures and humidity ratios. Using these values the sensible and latent loads are used to calculate final zone air temperatures, humidity ratios, and pressures.

Step 0. Virtual model realization

The first step in the model realization is the identification of building site and climate, by choosing a climate file. The correct exposure and geomorphological information should be verified and corrected. In the initial step the weather files of the nearest locality among those provided by the software library could be selected.

To insert the building geometry, the schematic of the building can be imported as *.dxf files. Subsequently, assigning the correct height, a virtual volume will be created through the extrusion of lines. The windows and door must now be inserted.

The set-points for systems have to be fixed: in the present study the values were 26°C for summer cooling and 20°C for winter heating.

Each envelope component has to be defined in the model by assigning for each layer the specific values of thermal conductivity λ (W/(m K)), specific heat capacity c (J/(kg K)), and density ρ (kg/m^3).

Subsequently the adjacency of the outermost surfaces have to be defined: for example, a ground floor slab should be recognized as surface laying on the ground.

Often it is not necessary to simulate the entire building (that could be studied entirely with a simpler software in steady state), being sufficient to represent only one apartment or a full vertical section from ground level

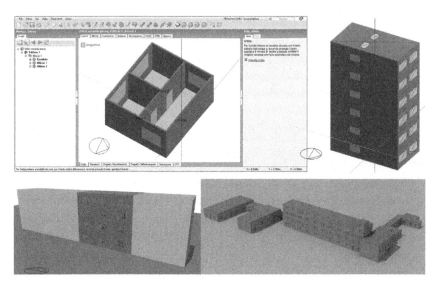

Figure B.5 Possible virtual models: a portion (or an apartment), a cross section with adiabatic walls, an entire volume.

to roof (thus including all levels) adopting lateral adiabatic surfaces (Fig. B.5). Indeed, the internal walls dividing two heated environment can be considered as adiabatic surfaces, since the two rooms are almost at the same temperature, and as a consequence the crossing fluxes can be assumed as negligible.

Step 1. Model calibration

The model should then be calibrated with measures. The calibration contemplates the tuning of a climatic input file with the monitored outdoor environmental conditions, obtained in the experimental surveys on the case study. The adopted approach [5] involves populating the model with occupancy values occurred in the monitored period. In the second stage the model must be further refined by identifying the most influential variables (ventilation schedules, infiltration based on blower door test) and adjusting the values through a reiterative process based on measured data. Several formatting steps were required to allow weather files to be used in EnergyPlus models, including generating ".stat" files using EnergyPlus weather statistics and conversion program.

So a virtual model that reproduces with a good approximation the observed values could be developed.

Table B.2 Example of errors calculation
Dining room, ground floor[a]

July 30—August 23	MBE$_{month}$	C$_v$ (RMSE)$_{month}$
Inside surface temperature	0.03	5.91
Inside air temperature	−1.00	5.81
Acceptable values	< ±10%	<30%

[a]Subhourly calibration type (10-min intervals).
Source: F. Stazi, E. Tomassoni, C. Di Perna, Super-insulated wooden envelopes in Mediterranean climate: summer overheating, thermal comfort optimization, environmental impact on an Italian case study, Energy Build. 138 (March 1, 2017) 716—732, ISSN 0378-7788.

The calibration was verified through the comparison of measured and simulated curves regarding the parameters (surface temperatures, air temperature, etc.). Moreover, two dimensionless indicators of error, mean bias error MBE and coefficient of variation of the root mean square error C$_v$ (RMSE), were calculated using formulae (B.19) and (B.20) as suggested by ASHRAE Guideline 14-2002 [6]:

$$MBE = \frac{\sum_{i=1}^{N_i}(M_i - S_i)}{\sum_{i=1}^{N_i} M_i} \tag{B.19}$$

$$C_v(RMSE) = \frac{\sqrt{\sum_{i=1}^{N_i} \frac{[(M_i-S_i)]^2}{N_i}}}{\frac{1}{N_i}\sum_{i=1}^{N_i} M_i} \tag{B.20}$$

where M_i and S_i are respectively measured and simulated data at instant i, and N_i is the count of the number of values used in the calculation. Table B.2 reports an example of calculation.

Step 2. Fixing standard conditions

This step is to ensure that the comparison between different envelope techniques is not influenced by the specific use of the heating system, or a specific profile of daily ventilation (as set in the calibrated virtual model). It involves the populating if the virtual model with standard profiles instead of personalized usage condition of the indoor environment.

In particular the internal gains profiles were fixed according to UNI TS 11300-1 [7] (Table B.3). The ventilation rate in winter was set at 0.3 air change rate per hour (ach), while the natural ventilation in summer was simulated with a continuous profile set to 1.5 or 4 ach depending on the cross-ventilation efficacy, respectively, referring to windows located in

Table B.3 Internal gains profiles according to UNI TS 11300-1

Days	Hours	Living room and kitchen (W/m²)	Other rooms (W/m²)
Monday−Friday	07.00 a.m.−05.00 p.m.	8.0	1.0
	05.00 a.m.−11.00 p.m.	20.0	1.0
	11.00 a.m.−07.00 p.m.	2.0	6.0
Saturday−Sunday	07.00 a.m.−05.00 p.m.	8.0	2.0
	05.00 a.m.−11.00 p.m.	20.0	4.0
	11.00 a.m.−07.00 p.m.	2.0	6.0

one facade or in two different facades (UNI 10375 [8]). Instead in the case of controlled mechanical ventilation the air change was set at 0.5 ach per hour according to UNI 10339:1995 [9].

The assumed set-points were 26°C for summer cooling and 20°C for heating.

Step 3. Parametric variations

Climates

Simulations were made on the dynamic model to provide consumptions and comfort levels in different seasons (assuming the introduction of a summer cooling system) and climate zones characterized by severe conditions. Bolzano (northern Italy) and London were selected for winter performance deepening, Palermo (southern Italy) and Cairo for a focus on the summer behavior.

For intermittent use of the system in mild climates a total use of 12 hours was assumed, while in all extreme climates a total use of 14 hours (as suggested by the standard [10]).

The heating/cooling season length for the various climate zones was fixed for Italian case studies referring to UNI/TS 11300-1 in which the length of the heating season is provided for each Italian climate zone.

For other climates (and also for Italian locations as comparison) the heating/cooling season was determined as the period during which it is necessary to activate the system to maintain an internal temperature not lower than 20°C in winter or not higher than 26°C in summer.

Such evaluations allow assuming an operating seasons as close as possible to the real use for every locality.

The heating/cooling season lengths finally adopted for the present study on the basis of the results obtained is reported in Table B.4.

Table B.4 Heating–cooling season lengths for the various climates simulated

Climate	Temperate climate		Winter extreme climates		Summer extreme climates	
	Ancona		**Bolzano**	**London**	**Palermo**	**Cairo**
	Csa^a: hot-dry summer Mediterranean climate		Dfb: cold temperate climate with warm summer	Cfb: humid temperate climate with warm summer	Csa: hot-summer Medit. climate near semiarid	BWh: hot desert climate
	Summer	**Winter**				
Begins[b]	June 1	November 1	October 15	September 15	May 15	April 15
Ends[c]	September 30	April 15	April 15	June 15	October 31	November 30
No. days	121	196	182	273	169	229

[a]Type of climate according to Köppen classification.
[b]Day of beginning of the heating/cooling seasons.
[c]Day of ending of the heating/cooling seasons.

Optimization scenarios and alternative modes of use

Different optimization scenarios were also evaluated. All the building components were varied according to the selected optimization strategy, thus introducing the same insulation levels not only on the vertical envelope but also on roof and ground floor slab. Moreover each variation was done by fixing all the other parameters to avoid results contamination. Fig. B.6 reports some parametric variation explored within the studies. The variations included both fictitious walls (as in the figure), to explore the effect of controlled variations, and real envelopes to deepen the behavior of technically feasible solutions.

For the intermittent heating/cooling, various time slots were considered to evaluate the impact of the mass with different operation profile in temperate climates. For all cases the heating system was switched on for a total of 12 hours per day, distributed according to the following time slots:
- 2 sections: 5.00 a.m.−9.00 a.m., 3.00 p.m.−11.00 p.m.;
- 3 sections: 5.00 a.m.−9.00 a.m., 12.00 a.m.−2.00 p.m., 5.00 p.m.−11.00 p.m.;

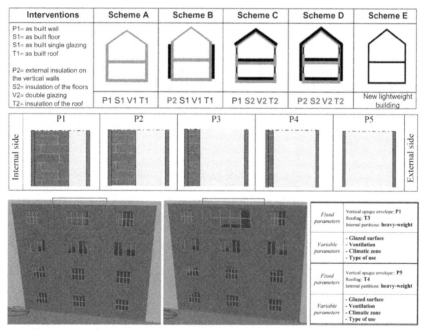

Figure B.6 Example of parametric variations on type of retrofit, type of external wall, glazed surface.

- 4 sections: 5.00 a.m.−9.00 a.m., 11.00 a.m.−1.00 p.m., 3.00 p.m.−5.00 p.m., 7.00 p.m.−11.00 p.m.;
- 4 sections night (only for summer cases): 2.00 a.m.−4.00 a.m., 7.00 a.m.−11.00 a.m., 2.00 p.m.−4.00 p.m., 7.00 p.m.−11.00 p.m.

Step 4. Simulation of passive solutions

The vented solutions were also simulated through EnergyPlus Airflow Network Tool. The Airflow Network model calculates the airflows based on AIRNET.

The vented cavities were simulated as a unique thermal zone with openings in the upper and lower parts.

In this cavity forced flow (due to pressure differences) and buoyancy flow (due to air density differences) coexist. The ventilation rate (q) through each opening in the model is calculated based on the pressure difference using wind and stack pressure effects:

$$q = C \times (\Delta P)^n$$

where q is the volumetric flow through the opening; C is the flow coefficient, related to the size of the opening; ΔP is the pressure difference between lower and upper openings; and n is the flow exponent varying between 0.5 for fully turbulent flow and 1.0 for fully laminar flow.

The wind pressure is determined by Bernoulli's equation, assuming no height change or pressure losses:

$$P_w = 0.5 \times \rho \times Cp \times v_z^2$$

where P_w is the wind surface pressure relative to static pressure in undisturbed flow (Pa), ρ is the air density (kg/m^3), v_z is the mean wind velocity at height z [m/s], and Cp is the wind pressure coefficient at a given position on the surface. It is a function of wind direction, position on the building surface, and side exposure.

Different types of ventilated facades were simulated by varying the insulation-mass positions and materials (Fig. B.7).

The solar wall (Trombe wall in its unvented use) was simulated with the algorithm "TrombeWall" for unvented cavities validated by Ellis [11]. It was modeled using standard EnergyPlus objects. A special Trombe zone is defined in the air gap between the Trombe wall and the glazing. The wall and glazing are standard EnergyPlus surfaces. The wall is connected to the main zone as an interior partition. The glazing is as a very large window covering the exterior wall of the zone. The critical difference

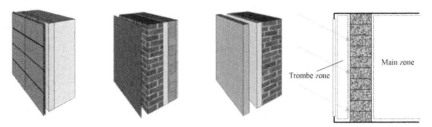

Figure B.7 Parametric studies on ventilated facades by changing the relative position between mass and insulation.

between a Trombe zone and a normal zone is the geometry: the Trombe zone has a much greater aspect ratio of height to width then a normal zone. The effect of such a high aspect ratio is to change the fundamental convective heat transfer phenomena in the zone. For this reason a special convection algorithm must be used for the Trombe zone. To introduce realistic input of ventilation rates due to Trombe walls in the model, experimental data about air velocity in the cavity were used as input.

REFERENCES

[1] US Department of Energy, EnergyPlus simulation software, Version 8.5.0 <http://apps1.eere.energy.gov/buildings/energyplus> weather data: <https://energyplus.net/weather>.

[2] Design Builder, Version 4.7.0.027. <http://www.designbuilder.co.uk>.

[3] U.S. Department of Energy, DOE Energy Plus 7.0 Input Output Reference: The Encyclopedic Reference to Energy Plus Input and Output, U.S. Department of Energy, 2011.

[4] G.N. Walton, AIRNET: A Computer Program for Building Airflow Network Modeling, Center for Building Research (U.S.), Edited by National Institute of Standards and Technology.

[5] M. Royapoor, T. Roskilly, Building model calibration using energy and environmental data, Energy Build. 94 (2015) 109–120.

[6] ASHRAE, Guideline 14-2002: Measurement of Energy and Demand Savings. ASHRAE, Atlanta, Georgia, 2002.

[7] UNI TS 11300-1/2:2008 (International standard ISO 13790:2008). Energy Performance of Buildings. Part 1: Calculation of Building Energy Use for Space Heating and Cooling. Part 2: Calculation of Primary Energy Use and of System Efficiencies for Space Heating and for Domestic Hot Water Production.

[8] UNI 10375:2011 (International standard ISO 13791:2005). Thermal Performance of Buildings—Calculation of Internal Temperatures of a Room in Summer Without Mechanical Cooling—General Criteria and Validation Procedures.

[9] UNI 10339:1995 (European standard EN 13779:2008). Air Conditioning Systems for Thermal Comfort in Buildings—General, Classifications and Requirements.

[10] D.P.R. 26 August n. 412: 1993. Regulations for the Design, Installation, Operation and Maintenance of Buildings Heating Systems in Buildings to Reduce Energy Consumption.

[11] P.G. Ellis, Development and validation of the unvented Trombe wall model in Energyplus (degree thesis for Master of Science in Mechanical Engineering), 2003.

INDEX

Note: Page numbers followed by "*f*" and "*t*" refer to figures and tables, respectively.

Printed in the United States
By Bookmasters